Advancing Collaborations for Water-Related Health Risk Communication

Awwa Research Foundation

About the Awwa Research Foundation

The Awwa Research Foundation (AwwaRF) is a member-supported, international, nonprofit organization that sponsors research to enable water utilities, public health agencies, and other professionals to provide safe and affordable drinking water to consumers.

The Foundation's mission is to advance the science of water to improve the quality of life. To achieve this mission, the Foundation sponsors studies on all aspects of drinking water, including supply and resources, treatment, monitoring and analysis, distribution, management, and health effects. Funding for research is provided primarily by subscription payments from approximately 1,000 utilities, consulting firms, and manufacturers in North America and abroad. Additional funding comes from collaborative partnerships with other national and international organizations, allowing for resources to be leveraged, expertise to be shared, and broad-based knowledge to be developed and disseminated. Government funding serves as a third source of research dollars.

From its headquarters in Denver, Colorado, the Foundation's staff directs and supports the efforts of more than 800 volunteers who serve on the board of trustees and various committees. These volunteers represent many facets of the water industry, and contribute their expertise to select and monitor research studies that benefit the entire drinking water community.

The results of research are disseminated through a number of channels, including reports, the Web site, conferences, and periodicals.

For subscribers, the Foundation serves as a cooperative program in which water suppliers unite to pool their resources. By applying Foundation research findings, these water suppliers can save substantial costs and stay on the leading edge of drinking water science and technology. Since its inception, AwwaRF has supplied the water community with more than $300 million in applied research.

More information about the Foundation and how to become a subscriber is available on the Web at **www.awwarf.org**.

Advancing Collaborations for Water-Related Health Risk Communication

Prepared by:
Rebecca Parkin, **Lisa Ragain**, and **Rebecca Bruhl**
Center for Risk Science and Public Health
The George Washington University Medical Center
2100 M St., NW
Washington, D.C. 20052

Heidi Deutsch
National Association of County and City Health Officials
1100 17th St., NW, Second Floor
Washington, D.C. 20036

and

Paula Wilborne-Davis
Association of Occupational and Environmental Health Clinics
1010 Vermont Ave., NW, #513
Washington, D.C. 20005

Jointly sponsored by:
Awwa Research Foundation
6666 West Quincy Avenue, Denver, CO 80235-3098

and

U.S. Environmental Protection Agency
Washington, D.C.

Published by:

Awwa Research Foundation

American Water Works Association

IWA Publishing

DISCLAIMER

This study was funded by the Awwa Research Foundation (AwwaRF) and the U.S. Environmental Protection Agency (USEPA) under Cooperative Agreement No. R829409-01. AwwaRF and USEPA assume no responsibility for the content of the research study reported in this publication or for the opinions or statements of fact expressed in the report. The mention of trade names for commercial products does not represent or imply the approval or endorsement of AwwaRF or USEPA. This report is presented solely for informational purposes.

CONTENTS

TABLES

FIGURES

FOREWORD

The Awwa Research Foundation (AwwaRF) is a nonprofit corporation that is dedicated to the implementation of a research effort to help utilities respond to regulatory requirements and traditional high-priority concerns of the industry. The research agenda is developed through a process of consultation with subscribers and drinking water professionals. Under the umbrella of a Strategic Research Plan, the Research Advisory Council prioritizes the suggested projects based upon current and future needs, applicability, and past work; the recommendations are forwarded to the Board of Trustees for final selection. The foundation also sponsors research projects through an unsolicited proposal process; the Collaborative Research, Research Applications, and Tailored Collaboration programs; and various joint research efforts with organizations such as the U.S. Environmental Protection Agency, the U.S. Bureau of Reclamation, and the Association of California Water Agencies.

This publication is a result of one of these sponsored studies, and it is hoped that its findings will be applied in communities throughout the world. The following report serves not only as a means of communicating the results of the water industry's centralized research program but also as a tool to enlist the further support of the nonmember utilities and individuals.

Projects are managed closely from their inception to the final report by the foundation's staff and large cadre of volunteers who willingly contribute their time and expertise. The foundation serves a planning and management function and awards contracts to other institutions such as water utilities, universities, and engineering firms. The funding for this research effort comes primarily from the Subscription Program, through which water utilities subscribe to the research program and make an annual payment proportionate to the volume of water they deliver and consultants and manufacturers subscribe based on their annual billings. The program offers a cost-effective and fair method for funding research in the public interest.

A broad spectrum of water supply issues is addressed by the foundation's research agenda: resources, treatment and operations, distribution and storage, water quality and analysis, toxicology, economics, and management. The ultimate purpose of the coordinated effort is to assist water suppliers to provide the highest possible quality of water economically and reliably. The true benefits are realized when the results are implemented at the utility level. The foundation's trustees are pleased to offer this publication as a contribution toward that end.

Walter J. Bishop
Chair, Board of Trustees
Awwa Research Foundation

Robert C. Renner, P.E.
Executive Director
Awwa Research Foundation

ACKNOWLEDGMENTS

The authors of this report are indebted to the following utilities, local public health agencies, and individuals for their kind cooperation and invaluable inputs to this project.

Water Utilities	City of Glendive Department of Public Works	Glendive, Montana
	Board of Water and Light	Lansing, Michigan
	New York City Department of Environmental Protection	New York City, New York
	Natural Resources Drinking Water Program	Redmond, Washington
	City of Tucson Water Department - Tucson Water	Tucson, Arizona
Public Health Agencies	Dawson County Department of Health	Glendive Montana
	Ingham County Department of Health	Lansing, Michigan
	New York City Department of Health and Mental Hygiene	New York City, New York
	King County Health Department	Seattle, Washington
	Pima County Health Department	Tucson, Arizona
Clinicians	All who volunteered to participate in either the telephone interviews or the national and regional workshops.	All of the above locations and members of the Association of Occupational and Environmental Clinics (AOEC)

We acknowledge the assistance of staff of the American Water Works Association (AWWA) who provided the list of water utilities for the survey components of this project, members of National Association of County and City Health Officials' (NACCHO) committees on Environmental Health and Prevention and on Bioterrorism for their invaluable suggestions during this project. We thank the staff of AWWA's *E-Waterworks* and AwwaRF's electronic posters for allowing us to notify utilities of our survey through their publications.

The authors thank the following individuals who made this project possible through their unique contributions, advice and insights.

- The members of the Project Advisory Committee (PAC), including Andrew DeGraca (San Francisco Public Utility Commission), Joyce Donohue (US Environmental Protection Agency), Caroline Johnson (Philadelphia Department of Public Health), Naomi Roseth (Australia Cooperative Research Center for Water Quality and Treatment), and Ron Wildermuth (Orange County Water District)
- Our project officer, Linda Reekie

The authors gratefully acknowledge the following staff members who provided critical assistance in completing this report:

- The George Washington University (GWU): Tee Guidotti, Bridget Ambrose, Kristen Facciolo, Michael Leong, and Polly Thibodeau
- NACCHO: Angela Ablorh-Odjidja, Cheryl Connelly, Librada Estrada, Tracy Johnson, Danielle Poux, Karen Roof, and Jessica Solomon

EXECUTIVE SUMMARY

Because water suppliers' are committed to providing excellent water quality to their customers, their efforts to continuously improve their processes and products based on the best available science will always be important. Advances in science and technology often raise issues and highlight potential threats to drinking water quality and utilities' credibility with their customers. The early involvement of public health officials and clinicians is essential for addressing these issues, and for building and maintaining effective collaborations before an emergency event. Utilities must be prepared to form new and maintain existing collaborations designed to address these current and emerging concerns. The purpose of this project was to develop strategies and tools that would enable water utilities, the public health community, and clinicians to collaborate and strengthen their capabilities for addressing water quality issues in normal and emergency conditions. The specific objective of this project was to design a Framework For Action with supporting materials to facilitate effective and dynamic partnerships for and beyond the utility service areas directly involved in this project.

This initiative included three major components: Data Collection, Data Analysis, and the Framework For Action. These components required gathering and assessing peer-reviewed literature and field data about existing communication capabilities, and the subsequent development, testing, and distribution of an evidence-based framework. The specific activities conducted were: literature reviews; interviews of utility, public health, and clinical personnel; an interdisciplinary workshop including presentations of case studies and group discussions; qualitative and quantitative data analyses; synthesis and interpretation of the data; preparation and field evaluation of a draft Framework; compilation of the lessons learned from testing the Framework; and finalization and dissemination of the Framework and the project's results.

The Data Collection and Analysis phases of the project provided the evidence on which the Framework For Action and the related Cost-Benefit Scorecard were based. No past and current three-way collaborations between water utilities, public health, and clinical medicine pertaining to collaborations for drinking water and health were found in the literature reviews. The field evidence from 10 jurisdictions indicated that local public health agencies (LPHAs) have good working relationships with water utilities. To obtain information about recent and existing communication collaborations, we conducted telephone surveys of 98 water utility and 160 public health agency personnel who were responsible for communication about water quality issues in the United States. Some of the most important survey findings were that utilities and clinicians expect LPHAs to serve as a communication bridge between them; this was found regardless of the size of community served or region of the nation. Although nearly all utilities and clinicians reported working with LPHAs, few reported working with each other on water-related health issues. Most utilities and LPHAs, like most clinicians and LPHAs, reported good to excellent working relationships with each other. The larger the population service area, the more likely the utility was to communicate with a LPHA at least weekly and address susceptible subpopulation issues with a health agency, but the less likely it was to designate a specific person to coordinate with the LPHA. Nearly all of the clinicians, who are specialists in environmental health, said that at least once a week their patients asked water-related questions, and the great majority of the questions were linked to recent mass media coverage. Nearly all of the clinicians said that they were willing to work with utilities; about half said they would do so on at least a quarterly basis.

To explore existing collaborations more fully, the five participating utilities (Glendive, Montana; Lansing, Michigan; New York City, New York; Redmond, Washington; and Tucson, Arizona) and their local partners met in nine focus groups during a two-day workshop convened in March 2004 in Washington, DC. All groups saw positive relationships – built slowly and consistently over time – as crucial to success in handling water-related health issues. Although these utility participants said their relationships with state and local public health agencies were generally good, they expressed frustration that agency staff could not interpret utility data. Agency staff said they wanted to know how to read water data reports. Utility and agency personnel view clinicians as lacking in essential knowledge about water systems and issues. This led them to conclude that clinicians are not prepared to answer their patients' questions. Clinicians view health agencies as credible vettors and communicators of water-health information, and thus rely on these agencies to provide them with the information they and their patients need. Agency personnel, however, expressed their concerns that they did not know what information clinicians need.

Based on the lessons learned from the surveys and workshop, a Framework was drafted to facilitate three-way collaborations (utility-health agency-clinical personnel). It was evaluated in each of the five participating regions during a one-day workshop, which included two hypothetical scenarios regarding different current and emerging drinking water-related health concerns mentioned by regional utility representatives. There were additional key findings from this series of workshops. Some regions had extensive lists of clinicians (usually physicians and not other clinicians) and long histories of working with clinicians to address public health issues. Although all utilities assumed that health agencies could readily communicate with clinicians, the participating health departments indicated that it was very challenging to keep their lists complete and fully updated. Further, most areas had not pre-tested or evaluated their water-health risk communication materials, and most had not obtained input from all three sectors (utility-health-medicine) to develop the materials. Participants found that their emergency response plans (ERPs) provided more effective means for handling acute water-health issues than chronic concerns; some areas said that their emergency drill procedures did not provide good means to capture "lessons learned" across organizations. Debriefings led by local personnel hindered the regions' abilities to improve their response capabilities. Further, participants in all areas recognized that they did not have accurate information about each others' roles and responsibilities or each others' organizational responsibilities for either routine or emergency conditions. For some, this incomplete knowledge could be consequential, particularly in rapid response environments. Further, participants in all areas revealed limited concepts of "clinicians," "stakeholders," and "risk communication." These conceptual limits prevented communities from identifying the full range of potential partners for addressing water-health risk issues.

The field exercises and literature review indicated several critical points for developing effective collaboration. Participants indicated that the key to sustaining relationships and effective decision-making is to repeatedly check on each other's assumptions. To build substantive partnerships, address misunderstandings as soon as they are known. Begin developing good working relationships so they are firmly in place before a crisis. However, change is difficult and challenging. As contexts and conditions change – whether social, political or organizational, the processes and goals of collaborations also have to change. Collaborators have to be willing to look at contexts larger than their own, think beyond what is familiar to them, and proactively involve communities and managers of key organizations. Good quality, tangible outcomes (e.g., plans, lists, skill building, etc.) strengthen partnerships, awareness, and capabilities for managing water-health concerns. Successful partnerships require local planning, design, management, mutually developed and realistic goals, and engaged institutions. Settings that favor collaboration include leaders who clearly dedicate time, make explicit commitments, contribute their expertise, and trust staff to pursue partnerships. Public relations and public information activities are a subset of the risk communication options needed to address public interests; using them alone, and not in a broader strategy, will be insufficient to provide effective, collaborative water-health risk communication services. Based on facts and not guesses about local issues and conditions, communication strategies must be rigorously tailored to suit local contexts, stakeholders and at-risk populations.

Two sets of tools – the Framework For Action and the related Cost-Benefit Scorecard – were developed based on this type of knowledge gained in the first two phases of the project. The underlying concepts of these tool sets for facilitating collaborations are: (1) complex issues – such as drinking water-related health issues – are easier to address through systematic approaches, and (2) tools designed to assist participants through the process, with tangible and useful outputs, support participants' efforts and reward their successes.

FRAMEWORK FOR ACTION

The Framework For Action was compiled based on "best practices" from the business literature and lessons learned in the data gathering components of the project. Fundamental principles that shape collaborations are systems thinking, strategic thinking, dialogue, shared vision, shared success, as well as shared responsibilities and risks. Creating an effective collaboration is not an all or none proposition; it is a process that matures over time. The Framework includes a seven-phase process with a series of tools to guide partners systematically for inventorying existing materials and authorities; fostering new insights and options for sustainable collaboration; identifying strategic issues; and ensuring feasible plans and effective mechanisms to continuously improve the partnership. These phases are titled:

- Prepare the process
- Start the dialogue
- Deepen the dialogue
- Learn from current conditions
- Create new ideas
- Plan to take action
- Take action

COST-BENEFIT SCORECARD

One aspect of effective collaborations for addressing water-health risks is having a structured means to identify the "best" risk communication option. This requires current information about the risk communication contexts, and available materials and resources. The Cost-Benefit Scorecard is a set of tools developed to guide the decision-making process for identifying the best option from a variety of possible approaches. Costs and benefits of different options may accrue to different stakeholder groups, so differences in impacts should be taken into consideration when choosing the best collaboration option. These tools also assist collaborators in identifying and considering these impacts.

The Scorecard includes three phases: clearly defining the risk communication problem, conducting a systematic analysis of the partners' resources and capabilities, and selecting the best option through a comparison of alternatives' qualitative costs and benefits. The steps in the Scorecard are titled:

- Problem definition
- Analysis
- Preparation for action

RECOMMENDATIONS

There is no "one-size-fits-all" method that will suit the conditions in all localities. Instead, each jurisdiction needs to build on the lessons learned by others, and design an approach appropriate for the resources and commitments available in their area. Many of the findings in this report point to recommendations that apply to all potential partners.

Applying the Framework or a similar process proactively is essential to position the partners for action under emergency or other stressful conditions. In a successful collaboration, each partner contributes something unique. The sum of their contributions will be greater than the collection of the parts; together their knowledge provides a more comprehensive view of conditions and more readily reveals major themes and trends.

Utilities should continue to explore approaches to broaden their identification of "clinicians" and "stakeholders," so potentially important supporters, particularly strategic partners, are not missed by approaching this task too narrowly. Set up opportunities for potential partners to exchange information about their roles, responsibilities, and services related to water-health concerns. Additionally, concerns about the limited lessons learned from emergency drill debriefing sessions need to be addressed. Do not assume that health agencies are able to contact all clinicians; strategize with health agencies and clinicians to determine how the gaps can be reduced. Recognize that risk communication materials need to be pre-tested and evaluated to ensure their effectiveness; agencies may be able to help with this. While LPHAs may be needed to interpret health data, their staff may need training to interpret water data reports.

Public health agencies are needed by both utilities and clinicians; however, they lack the resources to fully meet the expectations of these two groups. They should evaluate existing services and explore new strategies to build their resources and reach clinicians, especially those who serve populations at increased risk for water-related health risks. Community groups, including non-profits and libraries, may also be able to extend LPHAs reach. Agencies need to ensure that all of their organizational units involved with water-health issues are aware of each others' activities and services. Integrating water-health information in ongoing programs can provide added value at minimal cost.

Utilities, LPHAs, and patients expect clinicians to have updated information about water-health issues and to be able to answer patients' questions. Clinicians should know which sources of information are credible for their own education and which are suitable for their patients.

Where resources, time and the desire to collaborate are present, many actions can be taken to establish and strengthen collaborations for addressing drinking water-health issues. The tools and recommendations offered in this report provide many options for partners to deepen their working relationships. These are by no means the only possible methods to consider. We challenge our readers to start from where you are, use the tools that will serve your purposes, and proactively identify interested partners. Whenever you begin your efforts to build collaboration, you will be making progress toward improving the public's health and meeting their risk communication needs.

CHAPTER 1
INTRODUCTION

OVERVIEW

From ancient times, societies have taken steps to ensure the safety and adequacy of their water supplies (Rosen 1993). Community-level solutions were used worldwide, from the Orient to Europe. In Rome, both engineers and health officials worked to protect the public's health by providing safe water supplies. Over 150 years ago, the emergence of public health professionals from the field of medicine focused on critical public issues such as controlling infectious disease through safe drinking water. Technological developments that made this goal possible came from engineers dedicated to contributing to the public good. Although they approach issues in different ways and with different backgrounds, all three professions (engineers, clinicians and public health professionals) have long-standing values and activities focused on a common goal: protecting the public's health (Gorlin 1994, APHA 2001, Stevens 2001). The goal today is to revitalize that common commitment through new paradigms that will fit current conditions and needs.

The Institute of Medicine has called for stronger relationships among professionals and citizens to address public health issues (IOM 1988), but many societal forces have changed the context of public health, medical, and drinking water concerns. Increasing regulations resulting from the Amendments to the Safe Drinking Water Act (SDWA 1996) put considerable technological, economic, and communication pressures on water utilities. The Amendments also called for health care provider education and training about waterborne disease; this mandate led to the Health Care Provider Outreach and Education Working Group of the National Drinking Water Advisory Council (NDWAC). Although physicians are widely viewed by the public as highly trusted sources of information on drinking water quality and safety, for many reasons the public does not use physicians as sources very often (NEETF 1998 and 1999). Increasing access to the Internet has drastically changed the public's use of this information source for health topics.

In recent years, as the medical profession has struggled with the impacts of the managed care movement, many services traditionally provided by public health agencies were privatized, delegated to other agencies, or disbanded. Environmental health risk management strategies have shifted to more community-based dialogues, investigations, and joint problem solving (O'Connor and Gates 2000). Where used, these approaches have helped communities identify their shared values and resulted in new roles for public health agencies.

Surveillance of diseases, including waterborne diseases, is considered a key tool in public health, but has been a weak early warning system at best (LaPorte et al. 1993, Squires et al. 1998). Ways to improve surveillance systems, particularly with physicians' support, are being considered in dialogues about strengthening the public health infrastructure.* Improvements in data collection and interpretation will provide invaluable, timely information to water utility professionals and to communities at large, *if* effective risk communication collaborations are in place.

Events such as the 2001 attack on the World Trade Center have heightened concerns about civic safety and public health, stimulating increasingly earnest conversations and

---

* See the Centers for Disease Control and Prevention's Environmental Public Health Tracking Program at http://www.cdc.gov/nceh/tracking.

commitments to improve infrastructure and collaborations among engineers, utility professionals, physicians, and public health professionals. Commonly held motivations that underlie these dialogues about the public's interests exist but are not always explicitly stated. AwwaRF's RFP in 2002 calling for improved collaborations to address water-related health issues came at the ideal time to build on this momentum. Nationwide, there has been considerable public support and a sense of urgency to ensure that effective partnerships will be proactively designed and implemented for the public's good.

Our vision was that our unique partnership would increase the capabilities of water utilities, public health agencies, and clinical personnel to protect the public's health by addressing ongoing and acute water-related health issues. The three-phase project described in the report contained a set of science-based activities designed to deepen knowledge about the key elements for successful collaborations among water utility, public health agency, and clinical professionals. From this knowledge, we developed decision support tools to assist utilities and related health professionals in designing the collaborative mechanisms that will work best in their service areas.

This project included three major components: Data Collection, Data Analysis, and the development of tools to build collaborations for health risk communication (Table 1.1). The first two components focused on gathering and assessing field data about past and existing communication collaborations, and on potential collaborations in five different service areas of the United States. These areas ranged from small to very large cities located in different geographic zones; using surface, ground and reused water sources; and having different governance structures interacting with a variety of governmental organizations. The results of the Data Collection and Data Analysis phases provided the evidence base for developing the decision support tools: the Framework For Action and the Cost-Benefit Scorecard.

The specific activities completed in the project included literature reviews; interviews of utility, public health, and clinical personnel; an interdisciplinary workshop; qualitative and quantitative data analyses; synthesis and interpretation of the data; preparation and field evaluation of a draft Framework; compilation of the field lessons learned to finalize the Framework; preparation of the Scorecard to inform management decisions; development of recommendations for utilities, public health agencies, and clinicians; and dissemination of the project's results.

SIGNIFICANCE OF THE PROJECT

This project was based on scientifically sound literature, methods, and techniques from fields such as risk communication, risk perception, psychology, decision analysis, economics, public health and community health. The rich potential of the project was realized through the local knowledge, experiences, and insights provided by the participating utilities and health professionals. We convened a unique partnership of academic experts in risk sciences and community-based environmental health problem solving, and national organizations that represent the public health and clinical professionals most likely to be engaged with utilities on drinking water issues. Critical to the partnership were the five notably different utilities that offered exceptional opportunities to examine the commonalities and differences in collaborations that exist and may evolve in addressing water-related health risk issues. Additionally, we used a combination of complementary data collection techniques that yielded new information from which we drew key lessons for industry-wide application.

Table 1.1
Steps in building the framework

		Tasks								
Activities		Identify potential public health partners	Define existing utility-public health relationships	Identify potential variations in relationships	Evaluate issue-related effectiveness of public health personnel	Identify benefits, problems of working with these personnel	Select the best means for interactions	Assess the critical elements of successful collaborations	Develop a scheme to determine when to involve others	Test collaborative strategies and tools under different scenarios
Data collection	*Literature reviews*	X	X	X	X					
	Survey of utilities and health department personnel	X	X	X	X					
	Interviews of clinical personnel	X	X	X	X					
	Workshop of key parties in participating areas	X	X	X	X					
Data evaluation	*Identification of themes*				X	X		X	X	
	Assessment of communication needs				X	X				
	Description of communication alternatives				X	X		X	X	
	Cost-benefit examination					X			X	
Framework for action	*Draft framework*						X		X	
	Table-top exercises									X
	Final framework						X		X	
	Dissemination	X	X	X	X	X	X	X	X	X

OBJECTIVES

The purpose of this project was to develop strategies and tools that would enable water utilities, the public health community, and clinicians to collaborate and strengthen their capabilities for addressing water-health issues. Our specific objective was to design decision support tools and related guidance to facilitate effective and dynamic partnerships beyond the utility service areas directly involved in this project.

CHAPTER 2
REVIEW OF PAST AND CURRENT COLLABORATIONS

INTRODUCTION

The foundation of this project relied on the compilation and assessment of existing and new evidence from drinking water-related peer-reviewed studies, field experience, and risk communication and collaborations among utilities, health care providers, and public health officials.

Communication research has produced a variety of theories and frameworks for analyzing health risk communication and its influence on risk-reducing behavior.* To conduct meaningful research and communication activities, the underlying theory and methods used must match the level on which decisions are made about the risk. For example, the decision to drink water from the tap is made daily on an individual basis (not an organizational or societal level), so psychological theories and research approaches are appropriate. To complete this project, organizational and individual approaches and insights were appropriate and utilized at different points in the project.

The public health community has a long history of collaboration with both the water utility industry and clinicians. The success of many public health activities is dependent on clinicians' active involvement; e.g. in the conduct of surveillance, and many public education and information programs (CDC 2001). As mentioned in Chapter 1, public health and the water utility industry collaborated as a means of controlling infectious disease in the mid-to-late nineteenth century. About one hundred years later, the fluoridation of drinking water became a collaborative process between water utilities, public health organizations, pediatricians and dentists, which has successfully reduced the number of dental caries in the United States (CDC 1999).

Across the nation, state public health agencies (SPHAs) now regulate many water utilities. Public health practitioners, clinicians, and water utilities also collaborate on rule-making processes, such as the Disinfection By-Products Rule and the National Drinking Water Advisory Council established by the USEPA (USEPA 2003). LPHAs provide public services that address both health-related drinking water and recreational water issues; these are frequently conducted without utility involvement.

---

* There are three major categories of scientific theory that have been developed for risk perception and communication phenomena: cultural, organizational, and psychological (Krimsky and Golding 1992). Cultural theory focuses on the societal level, organizational approaches examine formal groups or entities within society, and psychological theories primarily address the individual scale and its fundamental importance to the larger levels.

LITERATURE REVIEWS

Methodology

Our literature reviews focused on capturing both research and practice-based evidence.* Documented reports of field experience were examined to identify key elements in successful collaborations, ascertain strengths and weaknesses, and develop recommendations and insights for the project and the drinking water industry.

Peer-Reviewed Literature

The peer-reviewed literature was identified primarily using automated search systems (e.g., Medline, Lexis-Nexis©, Aladdin, etc). We used structured queries and key word strings to survey a wide range of peer-reviewed literature. Google™ searches were used to ascertain any other literature that may have been omitted from traditional academic literature search sources. Personal contacts were made to identify unpublished and limited edition reports or other documented field evidence.

Field Evidence

Local public health agencies included in the field evidence search were identified using three methods. Members of the PAC for this project identified several specific jurisdictions to include in the data collection phase and recommended that five United States utilities with unfiltered water supplies also be contacted. Members of NACCHO's Environmental Health and Prevention and the Bioterrorism Advisory Committees represent a wide range of jurisdictions. Members whose jurisdictions have formal communication/collaboration plans with utilities were asked to provide both written documentation and, if needed, verbal clarifications. Snowballing techniques (e.g., identification of additional jurisdictions by contributing jurisdictions) were used to locate other sources of potential field collaborations.

The search for documentation was not limited to LPHAs; state health and/or environmental agencies and water utilities were also contacted to develop a better understanding of existing collaborative structures.

A systematic process was used to collect existing documentation. Unless a particular agency and/or person were identified in a specific jurisdiction, the LPHA was contacted by default. Phone numbers were identified using NACCHO's comprehensive database of LPHAs in the United States. Project team members asked to speak with a person responsible for water programs, specifically drinking water programs. Once the appropriate person was contacted, staff determined whether the agency had formal documentation of a relevant collaboration. Eligible evidence included Memorandums of Understanding (MOUs), Memorandums of Agreement, lawsuit settlement agreements, and/or communication plans.

---

* Results from scientific studies in the areas of risk perception and communication, health communication, psychology, decision analysis, sociology, public health, community health, environmental health risk management, and other fields were reviewed and synthesized for insights applicable to drinking water risk issues.

Such documentation was collected and reviewed by project staff to identify any points that needed clarification. If issues were found, the documentation provider was contacted for clarification. Key knowledge and recommendations were extracted from the full set of materials gathered.

Findings

Peer-Reviewed Literature

We found no analytical studies or directly relevant results describing community-based three way (utility, public health agency, and clinician) collaboration. Clearly, there is a paucity of peer-reviewed literature available on this topic. As we looked more broadly into general environmental health and medicine collaborations, however, we were able to develop some important insights.

Collaborations Between Public Health and Water Utilities. The cooperation of public health professionals and water utilities that began over 150 years ago has been maintained and developed as the significance of waterborne disease surveillance has continued to grow (CDC 2002). The relationship between public health practitioners and water utility professionals is crucial for the prevention and investigation of waterborne disease outbreaks (Ashford et al. 2003). Public health has identified multi-sector collaborations as critical for ensuring healthy communities and developing community based research programs, both of which must involve stakeholders (Israel 1998, NRC 1996a).

Collaborations Between Water Utilities and Clinical Medicine. One of the most recent efforts in collaboration and communication on drinking water and health has occurred between the AWWA, American College of Preventive Medicine (ACPM) and Dr. Patricia Meinhardt of the University of Rochester, with additional support from USEPA. Together, they have created a website, http://www.waterhealthconnection.org, that is an extensive resource capturing many aspects of water-related illnesses from a clinical perspective and presenting them in a textbook format. At the time of our project, the site was relatively new and no peer-review or evaluation of the content or usability of the site had been published. This website will be useful for some clinicians, but it is important to note that not all physicians or other health care providers use the Internet as their primary source of information (Lacher 2000).

Collaborations Between Water Utilities, Public Health, and Clinical Medicine. Fluoridation of drinking water, though still controversial, has served as a stimulus for collaborations between public health groups, drinking water utilities, physicians, and dentists (AWWA 2003, CDC 1999)

An emerging contaminant topic, cryptosporidiosis, was addressed through a large collaborative effort conducted by CDC and USEPA through the Working Group on Waterborne Cryptosporidiosis. This group resulted from a workshop hosted by CDC and USEPA in 1994 (CDC 1995), which in turn was convened as a result of the 1993 *Cryptosporidium* outbreak in Milwaukee. Public health professionals, water utility personnel, clinicians, and other stakeholders gathered to identify issues essential in addressing the threat of waterborne *Cryptosporidium*. The four areas were surveillance strategies and epidemiology study design; guidelines for a public health response, including prevention and outbreaks; prevention of cryptosporidiosis in susceptible populations, such as people living with HIV, and laboratory methods for environmental detection and diagnosis. The Working Group was formed to create guidelines for addressing the four areas. For almost two years, about one hundred people contributed to the final document, *Cryptosporidium and Water: A*

Public Health Handbook (CDC 1997). These guidelines have been in place for more than six years, yet no evaluations as to their effectiveness, utility, timeliness, or success in guideline implementation were in the peer-reviewed literature at the time of this project.

Several physician groups have been actively working with state or local public health agencies (S/LPHAs) and water utilities on drinking water issues. In 2000, Physicians for Social Responsibility created and distributed a guide on many aspects of drinking water and health (Balbus and Embrey 2000). The last section encourages readers to become involved and lists several public health and water utility related organizations. The release of the guide was accompanied by a workshop, which included not only health care providers from several fields, but water utility personnel and public health practitioners.

The ACPM has also worked on drinking water issues outside of their sponsorship of the waterhealthconnection.org website. One resource is an on-line lecture (http://www.acpm.org/ehealth/waterborne.htm) on what physicians need to know about drinking water and health (ACPM 2003). The presenters include water utility professionals and public health officials, as well as clinicians. There are no data available on physician utilization of this resource.

The effects of these collaborations are difficult to assess. There has been no published evaluation of these efforts. Other than fluoride, neither the short- nor long-term collaborations are known to have been sustained. It is likely that the collaborations described have educated specific groups of clinicians and developed collaborations on some level. The challenge is to work toward developing long-term, sustainable relationships that can reach out to the broader medical community.

Bioterrorism. Potential intentional threats to the safety of drinking water have come into the forefront of drinking water health concerns. Clinicians are vital in the detection of adverse health events (Ashford et al. 2003). Additionally, several of the agents of concern may be waterborne. The relationships between water utilities, public health officials, and clinicians, and their need for timely communications to combat such threats are not well recognized. Non-traditional surveillance methods may improve the timely detection of a bioterrorism event (Chang, Glynn, and Groseclose 2003). Water utilities may be able to use their expertise in monitoring to devise original approaches to public health surveillance.

ERPs and water security plans must involve not only public health officials and clinicians but also medical facilities such as hospitals and nursing homes. At the time of this project, no peer-reviewed literature was found on these topics. It is understandable that localities may not want to openly publish security-related materials; insights may need to come from other data collection activities.

Dr. Meinhardt and ACPM have developed a specialty website on bioterrorism and drinking water, *Physician Preparedness for Acts of Water Terrorism*, based on peer-reviewed sources. This site can only be accessed from http://www.waterhealthconnection.org. It uses the same textbook format noted above. No peer-reviewed evaluation of the site had been published by the time of our review.

Water Industry Communications. Communication tools designed to convey messages about water quality and related health issues have been in increasing use since the mid-1990s, when communication mandates were put in place (e.g., SDWA 1996). The best-known one-way tool is the Consumer Confidence Report (CCR), but utilities and state and local PHAs have developed other materials to provide additional information to the general public.

As required by the 1996 Reauthorization of the Safe Drinking Water Act, every United States water utility must annually report its water sources and water treatment process, contaminants detected in the drinking water and related health information. Specified language must be used in CCRs to warn people, such as immune-compromised individuals, who may be more vulnerable to waterborne contaminants. The required paragraph finishes with advising customers to contact their health care provider for further information on their specific health concerns as they relate to tap water use (USEPA 1998).

Studies have reported differing impacts of CCRs (Johnson 2000). One study showed that CCRs have little or no effect on customers' understanding of water-related topics (Trax and Snyder 1998 in Johnson 2000), while others indicate that CCRs have some positive influences on customers' confidence in water utilities (Odugbesan et al. 1998 in Johnson 2000, USEPA 2003, Lazo et al. 2004). The language of the CCRs is often difficult for utility customers to understand, if they are aware of the reports at all (Johnson 2000, Lazo et al. 2004). Because CCRs are neither entirely good nor bad, it is important to rigorously evaluate the assumptions on which CCRs are based and to learn how to improve them. However, most articles have focused on customers; we found no studies on how health care providers or public health personnel view CCRs for their own purposes.

Hyde, DeJong, and Ratzan (2000) examined the mandatory language of CCRs in focus group settings. Participating customers found the reports to be too dense; people just wanted to know if their water is "safe." Physicians participated in other focus groups; they felt their knowledge of drinking water issues and health was too sparse to adequately address patients' concerns.

One article offered physicians basic advice and background materials for counseling patients with questions regarding the CCRs (McCarty et al. 2000). No mention was made on whether or not the physicians knew about the CCRs or would feel comfortable in addressing patients' concerns. The authors listed related references and resources for health care providers. This article was found, however, in a journal with limited circulation.

CCRs may provide an opportunity to further link water utilities and clinicians. More research is needed as to the effect the reports have on patients bringing questions to their health care providers, and whether health care providers feel prepared to answer patients' questions.

Many utilities use fact sheets and media releases to directly notify both patients and health care providers about emerging issues. For example, as many utilities switched from using free chlorine to chloramines for drinking water disinfection, utilities made a concerted effort to notify patients on dialysis. The Nebraska Cooperative Extension (2002) prepared a fact sheet discussing the change from chlorine to chloramines, but at the time of our review (2003) no evaluation of this or other fact sheets had been published. There is a lack of information in the peer-reviewed literature regarding the effectiveness of materials and methods used by utilities to provide messages to the public.

Field Evidence

The jurisdictions contacted included Milwaukee, Maricopa County (AZ), Boston, Philadelphia, Seattle, Washtenaw County (MI), San Francisco, New York City, Portland, and New Jersey. Additionally, a MOU was received from Sydney, Australia (available at http://www.health.nsw.gov.au). Documentation of collaborations was identified in eight of these ten locales, but the collaborations vary in nature and subject matter; from task forces to formal, legal agreements and routine to emergency support services. Documents obtained for this project indicate that, where they exist, formal agreements tend to be completed at the state level (health or environmental agency) with utilities.

Boston has a communication agreement as a result of a USEPA-initiated lawsuit concerning unfiltered water systems. This agreement focuses on enhanced surveillance systems in which the utility contracts with the state health/environmental agency.

New York City has a communications agreement with the city health department. Under this Memorandum of Understanding, the city's Department of Environmental Protection supports seven to eight positions in the city health department; these positions are dedicated to syndromic surveillance, educational activities, routine sharing of data, and other water-related programs and issues (e.g., lead poisoning prevention, complaint response). Additionally, the city has developed plans for responding to water-related emergencies. The city health agency and utility staff are in close contact, to the extent of having – in addition to pagers – home and cell phone numbers of employees in case of emergency. Following the events of September 11, 2001, a general emergency response plan at the local level was expanded to include water issues.

Beginning with poor communication between agencies and with the public, the Philadelphia (Pennsylvania) LPHA and water utility began to realize that, should an event like the 1993 cryptosporidiosis outbreak in Milwaukee occur in their area, they were unprepared to collaborate with each other. As a result, the utility began to provide water quality data to the health department, engage customers more actively, identify spokespeople, and create a Water Quality Committee that consisted of the health department, physicians (including those who work with vulnerable populations), academicians, regulators, and other stakeholders. The LPHA established cryptosporidiosis as a reportable disease, implemented active surveillance for diarrheal illness, shared surveillance data with the utility, and developed improved communication technology. The Water Quality Committee developed a microbial communication plan with objectives to describe parameters and standard surveillance procedures for water quality and waterborne diseases, establish threshold levels for communication triggers, and describe the plan in terms of the utility's operations. The Committee evolved over time (see Appendix A) and continues to work on issues of water quality and surveillance data; it makes technical decisions, serves as a communication portal to others, and plans for emerging issues.

In San Francisco, the utility supports two full-time positions for epidemiologists to conduct disease surveillance and provide informational support related to waterborne contaminants. For example, they post fact sheets on the web, develop Positive Response Plans (e.g., for *Cryptosporidium*), and provide consultations during emergency events.

The State of New Jersey developed a tax on water utilities to create a drinking water quality institute designed to review federal drinking water standards and determine whether state regulations needed to be more rigorous. The institute includes representatives of both the state health department and utilities to balance public health and utilities' feasibility concerns.

In Portland, if a violation or other emergency situation occurs that poses an acute risk to human health, a public notice must be provided to all persons residing within that jurisdiction and served by the water system. The city has a specific plan to address health issues associated with lead in drinking water, and a general plan for boil water alerts (BWAs) but not for other emergencies. Portland's BWA communication plan is based on USEPA's public notification regulations, while the state water bureau coordinates the communication plan as needed beyond the city's jurisdiction. The bureau has established a comprehensive emergency management program that utilizes the incident command system to manage emergency events, including those involving water. Under the program, the Incident Commander has overall responsibility for managing the incident and is ultimately responsible for making the decision to issue a boil water notice. Prior to public delivery of any water emergency notification, the public information officer, water quality and treatment manager, regulatory compliance officer(s), members of the state's Health Division, and the County Health Officer must be notified.

In a unique instance, legislation in Arizona called for the formation of a task force after a cryptosporidiosis outbreak in another state. This task force includes representatives from the Maricopa County and state health departments, Department of Natural Resources, the utility, the department of public works, and others. The task force meets regularly to discuss varying water issues and has created an ERP should the water system be compromised.

Finally, Sydney's health department and water utility have a formal Memorandum of Understanding that describes their mutual responsibilities to protect the public's health as it relates to drinking water quality. The Agreement calls for sharing of strategies, long-term and annual plans, monitoring data and information, and the provision of access, consultation and advice during routine as well as emergency conditions. Another part of the Agreement states that both organizations will jointly provide health education services for the public.

From the literature gathered, the following issues were identified as needing further investigation.

- Who at the S/LPHA is the liaison to the utility?
- Does the type of formal plan have any bearing on the quality of the collaboration between the utility and the S/LPHA?
- To what extent were the events of September 11, 2001, responsible for stimulating collaborations between unique partners, including local public health and utility communities? Does this origin limit the potential for sustained collaboration?
- Do state health and/or environmental agencies' collaborations with utilities bypass other collaborations at the local level or do state agency-utility partnerships enhance them?

During NACCHO's annual meeting in 2003, NACCHO staff obtained information from additional jurisdictions about their health agency-water utility collaborations. From these efforts, interactions between the LPHAs and utilities were typically found to have the following traits: *cooperation*, *cultural relevance*, *longevity*, *trust*, *communication*, and *need*, noting that each entity needs the other and neither is superior. The health officials felt that regular communication was key to a good working relationship. Thus, when a crisis situation does occur, communicators already know each other, and the mandatory communication is straightforward and less complicated. Project staff recognized that while trust is built over time, through performance, there is always a need to "self-check." Longevity comes into play when there are staff changes. If the relationship is already established and strong, a change in staff should ideally have little impact. The discussants said that, overall, relationships tend to be sustained over time, except when one person negatively changes the entire relationship between agencies. Examples were offered that had impacts from the 1970s through early 1990s.

In general, the field evidence indicates that LPHAs have good working relationships with related water utilities. Whether the communication is mandated, voluntary, formal, or informal, the overall finding is that communication *is* taking place at the local level, although it may not be readily apparent or characterized. However, an additional challenge is faced by larger organizations; e.g., ensuring that all units (including but not limited to maternal and child health, community health, epidemiology, communicable disease, HIV/AIDS, and health services) that may have an interest in a water-related health issue are aware, appropriately informed and engaged. From this basis, we designed surveys to assess the nature of working relationships between utilities, LPHAs, and clinicians.

INTERVIEWS

Methodology

Both individual and group interview methods were used to gather information and insights into existing and potential utility-health collaborations focused on drinking water-related health risk communication. Brief telephone interviews were used to assess broadly *what* utility, LPHA and clinical personnel had experienced when interacting with each other about water-related health issues. Extensive, facilitated group discussion methods (focus groups and tabletop exercises) were used to deepen knowledge about the nature of collaborations in five specific areas and the reasons for their successes and limitations. The primary aim of the group interview activities was to assess *why* collaborations did or did not occur. Consequently, the data collection methods were complementary; e.g., the surveys were designed to obtain limited information on a broad geographic basis, while the group methods were used to deepen knowledge in a few, differing types of utilities in urban and rural areas of the United States.

Surveys of Water Utility and Local Public Health Agency Personnel

Two national surveys were conducted to obtain information from water utility and public health professionals practicing in a wide range of settings.* This method was used to broaden and deepen knowledge, beyond what could be obtained from literature reviews, about past and existing collaborations for water-related health risk communications.

The survey instruments for the utilities and health agencies were very similar, essentially only reversing "utility" and "health agency" in the questionnaire (see Appendix B). Participants were asked not only about their communication and collaboration experiences, but also their current views and needs related to future collaboration. Furthermore, we investigated what issues, interests, and priorities these professionals did and did not have in common. A few questions were unique, but similarity overall was important to facilitate comparisons across the two groups.

There was no literature available to estimate the percentage of professionals who have any particular view related to utility-health collaboration. Based on expert model research (Morgan et al. 2002), however, it is not unreasonable to expect that any one view may be held by 10% of the survey participants. Starting with this assumption and a study population of 3,000 organizations, we estimated that we could predict the true prevalence of any particular trait or view with a 99% confidence interval if we obtained 100 complete interviews.

Water utility and local public health agency representatives were identified using standardized, random number-based selection processes. We obtained a list of water utilities from AWWA and a list of health officials from NACCHO. There were 2,852 county/city public health agencies and 4,428 water utilities available from these sources. The NACCHO membership list was

* The questionnaires were reviewed and approved by GWU's Institutional Review Board (IRB) before initiating any efforts to recruit participants for these interviews (IRB #U070222).

Table 2.1
Number and percentages of NACCHO members by region (2003)

Region	Number	Percentage of all members
Northeast	672	23.6
Southeast	591	20.7
Midwest	1,210	42.5
West	377	13.2
Total	2,850	100.0

Table 2.2
Number and percentages of AWWA members by region and system size (2003)

	Small		Medium		Large		Totals	
Region	No.	(%)	No.	(%)	No.	(%)	No.	(%)
Northeast	633	(14.3)	82	(1.9)	14	(0.3)	729	(16.9)
Southeast	641	(14.5)	122	(2.8)	17	(0.4)	781	(17.7)
Midwest	1481	(33.5)	136	(3.1)	25	(0.6)	1642	(37.2)
West	1078	(24.4)	172	(3.9)	18	(0.4)	1268	(28.6)
Totals	3833	(86.5)	512	(11.7)	74	(1.7)	4419	(100.0)

divided into four regions* and the percentage of LPHAs in each region was calculated (Table 2.1). Agencies were then randomly drawn within each region to create a proportional, random sample for the nation. There were 1,500 agencies in this sample selection.

Similarly, the AWWA membership list was divided by region and size of the utility (small, medium, large†) based on the number of service line connections (Table 2.2). The percentage of utilities by size was calculated for each region. Utilities in each category were then randomized to create a proportional random sample of 1,500.

---

* Northeast: Maine, Vermont, New Hampshire, Massachusetts, Rhode Island, Connecticut, New York, New Jersey, Pennsylvania, Maryland and Delaware

Southeast: District of Columbia, Virginia, West Virginia, North Carolina, South Carolina, Georgia, Florida, Alabama, Mississippi, Louisiana, Tennessee, Kentucky, Arkansas

Central: Ohio, Missouri, Indiana, Illinois, Michigan, Iowa, Kansas, Texas, Oklahoma, Nebraska, Wisconsin, Minnesota, North Dakota, South Dakota

West: Montana, Wyoming, Colorado, New Mexico, Arizona, Utah, Idaho, Washington, Oregon, California, Alaska, Hawaii

† The sizes were defined as follows: small is equal to or less than 10,000 connections; medium is 10,000 to 99,999 connections, and large is 100,000 or more connections.

Table 2.3
Number and percentages of eligible organizations by type and study status

Study status	Water utilities No.	(%)	Local public health agencies No.	(%)
Eligible to participate	4,428	(100.0)	2,852	(100.0)
Contacted to participate	*3,000*	(68.0) *	*2,852*	(100.0) *
Volunteered to participate	*126*	(4.2)†	*209*	(7.3)†
Unable to reach by telephone	3	(2.4)	9	(4.3)
Telephone contact completed	*123*	(97.6)‡	*200*	(95.7)‡
Refused	8	(6.5)	1	(0.5)
Lost to follow up	5	(4.1)	14	(7.0)
Miscellaneous losses	12	(9.8)	25	(12.5)
Participated	98	(79.6)	160	(80.0)

* The basis for the following row, "Volunteered to participate."
† The basis for the following two rows, "Telephone contact completed."
‡ The basis for the percentages shown below this row.

In December 2003 through February 2004, letters of invitation were sent to the water utilities and health agencies drawn for the first samples. Additionally, AWWA's *E-Waterweek* and AwwaRF and NACCHO electronic postings were used to recruit volunteers.* The final samples are shown in Table 2.3.

Although the volunteer rates among those eligible were far lower than anticipated, the completion rate among volunteers who could be contacted was good (about 80%). Because water utility and public health agency personnel receive a large number of surveys, this may have contributed to the relatively low response rate. Further, from a few callers requesting information, we believe that decisions to not respond may have been strongly influenced by federal agency and others' cautions against providing information to unknown parties that may affect water security. Although callers were satisfied that the survey was legitimate and would not compromise their security, we can not estimate how many people had such concerns but never called the number provided in the letter of invitation to verify the credibility of the survey. Additionally, the initial mailing response rate may also have been affected by the year-end holiday season backing up mail past the original response date. Finally, errors in the mailing list contact information resulted in about 80 letters being non-deliverable. Reasons noted on the returned mail included incorrect addresses, retirement, and personnel changes.

---

* OpinionSearch, a Canadian-based social research services and field data collection company, conducted the 15-minute interviews using well-trained interviewers and computer-assisted technologies. No interview was conducted until the potential participant was informed about the survey's purposes, sponsorship, and privacy and confidentiality protections, and until the participant had the opportunity to have their questions about the survey addressed. Additionally, no interview was conducted without the participant's explicit consent, and understanding that s/he may stop the interview or refuse to answer any question at any time.

While much of the analysis relied on descriptive statistics, analytic methods such as correlations and tests of significance (chi squares and confidence intervals) were used to the extent appropriate. We used Microsoft Excel and Access to manage the database and SAS® programming to analyze the data.

Interviews of Clinicians

Our goal was to gain insights into the issues, interests, priorities, and problems that clinicians have when involved in or considering drinking water-health communication issues. AOEC has a nationwide membership of 60 occupational and environmental medicine clinics along with more than 300 individual members. The membership includes physicians, nurses, industrial hygienists and other health professionals with interests in occupational and environmental health. We anticipated recruiting about 50 clinicians for 15-minute focused interviews, scheduled at a time convenient for each clinician. However, multiple mechanisms (e.g., e-mail notices and newsletters, etc.) used by AOEC did not yield the number of volunteers anticipated. A trained and experienced interviewer at GWU contacted the 38 volunteers during February and March 2004.*

Thirty interviews were completed; non-participants were sick, unavailable, or had moved at the time of the study; two volunteers were found to be ineligible because they did not see patients. The rate of completed interviews among the clinicians (79%) was comparable to that for the survey participants (79-80%). Our clinician interview instrument is included in Appendix B.

Findings

Surveys of Water Utility and Local Public Health Agency Personnel

Survey participants worked in either water utility or local/county public health agencies (LPHA). The participants tended to be male managers who had worked in their profession for over 20 years. Additional demographic information is detailed in Appendix C (Tables C.1 and C.2).

We found that utility respondents contact S/LPHAs more often than LPHAs contact utilities. In addition, most respondents for both utilities and LPHAs indicated that their organization has an official designated for communicating with the S/LPHA or utility. Few said that their organization had a formal agreement with the S/LPHA or utility for drinking water communications, though a large majority of all survey participants said their organization has worked with a S/LPHA or utility in the last five years, often on boil water alerts and/or emergency response plans. See Appendix C, Table C.3 for more information on the survey results.

Utility and LPHA personnel also answered questions about their relationships with health care providers (HCPs) (see Appendix C, Table C.4). More than half of LPHAs – but relatively few utilities – report ever working with HCPs on drinking water issues. Survey results indicated that

* The participants were a sample of convenience and may not represent the full range of views held by all AOEC physicians or all clinicians at large. The interview protocol was reviewed and approved by the GWU IRB before initiating any efforts to recruit participants for these interviews (IRB #U070222). No human research or data collection can be conducted without prior approval by the IRB. As for the surveys (above), we complied with the IRB's requirements to fully inform participants about the study and their rights before any questions were asked. We did not maintain any records that could be used to identify the participants and kept all interview data in a locked file in the PI's office.

LPHAs communicate with HCPs more frequently than utilities. Most respondents said that they communicate with HCPs once per year or less; physicians, nurses, and dentists were the most common HCPs with whom they communicated. A majority of utility and LPHA participants described their relationship with HCPs as excellent or good.

Analyses were conducted to explore whether certain characteristics of water utilities or LPHAs influenced their interactions with utility personnel and LPHA staff. Although the sample size limited statistical power, several significant associations were identified (see Table C.5 in Appendix C). For example, the larger the population (over 100,000) in the area served, the more likely the agency or utility was to communicate with the other party weekly or more often, to have ever worked with the other party on drinking water communications, and to have ever worked with the other party on water security-related ERPs. Managers* at water utilities were more likely than other utility respondents to report that they had ever worked with a SPHA, had collaborated with a S/LPHA on a drinking water communication in the past five years, and had recently participated in a joint communication activity to produce a CCR. In addition, participants who had worked in the water utility industry for over ten years reported that the best ways to improve their relationships with LPHAs were through more collaboration and increasing the frequency of their communications. Newer employees emphasized the need to make and build contacts in LPHAs by developing a better understanding of roles and better interpersonal relationships.

Interviews of Clinicians

The 30 participants in the telephone interviews tended to be male, physicians, who had been in practice for over 10 years (see Table C.6 in Appendix C). Key findings from these interviews are described in Tables C.4 and C.7. Generally, clinicians are more likely to contact S/LPHAs than utilities about drinking water concerns. Few had ever been contacted by either a utility or S/LPHA about collaborating on a drinking water issue. Of the few clinicians who had collaborated with an LPHA on a drinking water issue, they generally viewed the collaboration as clinically relevant and successful or very successful. Education was the joint activity most often stated as successful; responses to emergency events were least successful. Almost all clinicians said they were willing to collaborate on drinking water issues. Half said they wanted more frequent communication or more collaboration with S/LPHAs.

Nearly all clinicians said their patients asked questions about drinking water issues, and these questions generally came daily to monthly. Newspaper articles and television stories were often cited as prompting the patient's question. Most clinicians had diagnosed a disease or condition that they judged was water-related; usually patient symptoms prompted them to look for water-related etiologies.

To find water-health information, generally clinicians used online tools such as Medline or turned to the Centers for Disease Control and Prevention (CDC), the federal Agency for Toxic Substances and Disease Registry, or medical journals. Less often, they turned to LPHAs, USEPA, and water utilities. About half of clinicians remembered receiving information about drinking water issues from the S/LPHA and/or the water utility in the last five years. Many clinicians

---

* These same relationships were generally true with employees who had worked in the profession a long time. This would be expected since the rank of manager is related to the length of service in the profession.

indicated that they want more information about a number of specific microorganisms and chemicals (see Table C.6 in Appendix C for more information).

Lessons Learned

The major findings from the surveys and the clinician interviews were:

- More utilities than LPHAs reported working with each other on a drinking water issue.
- A large majority of utilities and LPHAs stated that their relationships were excellent or good.
- More LPHAs than utilities had ever worked with clinicians on water-health issues.
- More LPHAs than utilities had worked with clinicians on susceptible subpopulation issues.
- The larger the population service area, the more likely the utility was to communicate with a S/LPHA, work with a S/LPHA to address susceptible subpopulation issues and prepare a written communication, but the less likely the utility was to designate a specific person to coordinate with the S/LPHA.
- The LPHAs that had a large number of utilities in their area were more likely to communicate with a utility at least once a week.
- Utility staff members who had been in the industry for over ten years said that they would like more collaboration and communication with S/LPHAs.
- Nearly all of the clinicians reported that their patients asked water-related questions; clinicians related the great majority of the questions to mass media coverage.
- Clinicians viewed their relationships with LPHAs as excellent or good.
- Although few had ever worked with utilities, nearly all of the clinicians said that they were willing to work with utilities.

GROUP DISCUSSIONS

Methodology

One data gathering activity that generated deeper information than could be obtained in brief telephone interviews was the two-day intensive workshop, held on March 19-20, 2004, in Washington, DC. This activity focused on existing and desired collaborations. The outcome of the discussions produced collective observations about current practices and suggestions for improvements. The other group data gathering approach involved visiting each participating region and conducting tabletop exercises to assess opportunities and limitations of potential collaborations; these are described in Chapter 3.

The workshop involved the entire project team (GWU, NACCHO, and AOEC), the Project Advisory Committee (PAC) members, the Project Officer, and public health, medical and water utility personnel from Glendive, Montana; Lansing, Michigan; New York City, New York; Redmond, Washington; and Tucson, Arizona. We invited the regional participants to identify three other key stakeholders from their jurisdiction to attend the workshop. This activity involved background presentations, case studies, and guided group discussions – typically known as focus groups – to collect information about existing and desired utility-water collaborations for improving water-health risk communication purposes. At the conclusion of the workshop, each region determined what actions they would take to move toward effective collaboration.

At the beginning of the workshop, we provided background information and clarification of the workshop goals. We provided the results from prior, similar GWU-lead workshops,* a brief summary of our literature review and preliminary survey results as background information to the participants in this project. PAC members were invited to present their experiences with utility-health collaboration; these provided the participants with examples to refer to in their group discussions.

Based on standard methods, we conducted focus groups – one each for utility representatives, LPHA representatives, clinicians, and others (see below) – to encourage participants to share experiences with their peers, help them identify their collective issues and needs about water-health communication collaborations, and facilitate their thinking before meeting with their colleagues from the other groups (Rice and Ezzy 1999). The themes from each group were presented to all other participants and discussed to synthesize the results. Next, we convened the participants by each of the five regions, so they could compile their experiences, identify the lessons they had learned, and develop recommendations to share with the other regions. We sought their evaluation of the advantages and disadvantages (including cost implications) of the communication approaches they had used and wanted to use.†

We systematically examined the workshop notes and located common concepts and subconcepts (Ritchie and Lewis 2003). We used an iterative process involving the workshop facilitators to develop the final themes from the data collected. Further we presented the workshop results to the regional exercise participants and obtained their comments to finalize the summary presented here.

Findings

Nine focus groups were hosted during the two-day workshop. Groups ranged in size from three to twelve participants; all individuals participated actively in the group discussions. Facilitators guided the groups' discussions using a set of questions, but not all questions were asked in all groups. Some groups chose to concentrate on specific issues. These variations in coverage account, in part, for the differences in the group descriptions below.

Discipline-Specific Groups

Four discipline-specific focus groups were convened: water utilities, public health agencies, clinicians, and other stakeholders (e.g., academicians and elected officials). Each met for about one and a half hours on the first day of the workshop.

One key theme that emerged is that all groups saw positive relationships as crucial to success in handling water-related health issues. Many also stated that these relationships need to be built slowly and consistently over time; they should not be expected to just happen during a

* GWU hosted a similar workshop in fall 2002 specifically focused on evaluating risk communication tools for water-related health risk issues under our Cooperative Agreement with the USEPA Office of Water.

† As for the surveys and interviews (above), the focus group sessions also required and received the review and approval of GWU's IRB (IRB #U070222).

crisis. Participants also stressed that problems need be to be addressed upfront with a maximum degree of trust and credibility. Similarly, all groups indicated that they felt the public had some growing concerns about water issues. Participants said that these concerns may result from the public seeing confusion rather than clear leadership during emergency water events.

Utilities. The participants said that their working relationships with public health agencies were generally good, but that they varied. The variation was particularly apparent when the participants considered regulatory vs. non-regulatory communications; e.g., at times rules and regulations impede effective communications. "Regulatory" communications encompassed reporting, consultation on CCRs, and monitoring issues within the jurisdictions of the utilities and S/LPHAs. These communications were characterized as routine, manageable, and regular. The second type of communications included activities outside of regulatory requirements. Examples included the development of educational materials, program development, or other types of health consultations. This type of communication was characterized as transient, more challenging, and dependent on previous relationships and knowledge about both utility and S/LPHA abilities and limitations.

Utility personnel stated that they need different things from S/LPHAs, depending on whether or not the S/LPHA was the regulating agency or not. When the LPHA was not the regulator, utilities tended to focus their interactions on the SPHA. More generally, the participants expressed their frustration with S/LPHA staff not reading water quality reports and/or not being able to interpret the reports. Above all, however, utility personnel felt that it was important to have a long-term, substantive relationship with the S/LPHA staff with whom they had to work. From that base and an understanding of each other's interests and needs, utility personnel anticipated being able to work with S/LPHAs to address many problems.

In contrast, the utilities had very limited relationships with clinicians and did not see the need for more active relationships. The participants did not know what clinicians might want from utilities; they want clinicians to tell them what they need. They expect to interact with clinicians through the S/LPHAs. Utility representatives indicated that they might contact a clinician about a specific issue (e.g., fluoride), but want customers to take their health questions to their clinicians and not the utility. The participants stated that clinicians have little knowledge of water, water treatment or water systems, and therefore may not be able to address patients' concerns. The participants expressed their desire for clinicians to learn more about water-related issues. Key quotes characterizing points made by utility representatives are shown in Table D.1 of Appendix D.

Local Public Health Agencies. The LPHA representatives stated that personal relationships with utility staff were crucial; they saw protocols for operations as secondary to relationships. They said it is very important to know the specific person in the utility with whom they must work to address issues. Representatives also indicated that they typically meet with utility staff quarterly or biweekly. Based on their interactions, LPHA personnel felt they need to know more about utilities, particularly how to interpret utility data. However, health agency personnel acknowledged that there are many utilities for them to know about, so they tend to focus on the biggest one in their areas. In one case, a utility had funded a position in the LPHA to increase communications. In other cases, the utility's Public Information Officer (PIO) was the individual that LPHAs contacted. The participants stated that they knew PIOs were expected to return calls the same day and therefore expected them to serve as the bridges to the right person in the utility who could meet their needs. Health staff also stated that they trust utilities because they self-monitor, but also noted that utilities have an interest in not seeing problems.

Health agency representatives said that they have strong relationships with clinicians, although contact may not be frequent except in emergencies. They stated that an important role they have is to be a credible source of information for clinicians, and in turn clinicians need to be a credible source for their patients. Health staff readily noted that water competes with many other issues in a clinician's agenda and in the LPHA-clinician relationship. LPHA staff said clinicians do not know disease-reporting regulations, and not every clinician needs to know everything. It is particularly difficult to reach some clinicians, especially those in small or ethnic communities and those without hospital affiliations. Some of these difficulties have been exacerbated by the fragmentation of the health care system. When relationships need to be built, typically the director of the LPHA takes the lead. However, agencies have learned that not all clinicians want to have a relationship with them.

Health agency personnel also discussed their relationships with the press, public, academicians, and elected officials. They noted that there are huge health disparities among the populations they serve, and that ethnic populations often do not trust S/LPHAs or governments. Maintaining trust is crucial when addressing water and health issues. Some expressed concerns that water issues often do not reach restaurant or school staff. Concerns were raised that protocols may need to be developed to address these stakeholders. It is important to reduce people's misconceptions that water is sterile; this could be done through a personally staffed call-in center, school education programs, and the media. Although relationships with the media vary, LPHA staff members know that they must be proactive with media personnel. Building early, trusting relationships is important, so that – when water is a priority – that connection is already firmly in place. However, LPHA representatives noted that utilities see LPHA relationships with the media as potentially problematic. Additionally, they recognized that the loss of public trust through the mismanagement of water-related health issues would require many years to heal, if ever. In some cases, it may be more effective for community representatives to reach the public about water-health concerns.

LPHA personnel expressed more variable experience with elected officials. They characterized their relationships with them as very challenging but "we have to work together." LPHA staff members see their role as providing information to elected officials; thereby helping them make decisions and announce those decisions to the public. During a crisis may be the only time that LPHA staff members interact with elected personnel. Few of the participants expressed much knowledge of the role of their local Board of Health.

LPHAs see academic centers as resources for information and emergency support. Although they wanted them to be strong, many LPHA representatives said that their relationships with academicians were either not strong or non-existent. Some indicated that hospitals had links to academic centers, or that the LPHA related to educational institutions on specific issues.

As they reflected on their own needs, LPHA personnel stated that they need to know more about specific contaminants, crises, water supply systems, and what communities need to know. LPHAs would also like to involve clinicians in drills, while putting water in perspective to other issues.

LPHA staff felt that what facilitates relationships is knowing who to talk to, increasing their own knowledge about utilities and water systems, and assisting utilities in increasing their knowledge of health agencies and the health aspects of water issues. Selected quotes illustrating major points made by LPHA staff are given in Table D.2 of Appendix D.

Clinicians. Overall, the participating clinicians trusted government, health agencies, and water utilities to do the right things to protect their patients' health. Representatives stated that they had little experience with water utilities, but would like utilities or health agencies to take the initiative to reach out to them using one of their more scientific staff (e.g., a clinician, scientist, or engineer). Some said they would like to develop appropriate and patient-friendly literature in collaboration with health agencies and utilities, but others said they do not have time for this. Others pointed out that they want only simple e-mails; that is, none that include lengthy, slow-loading enclosures. In general, the participants felt that food-borne pathogens were more important in their practices than waterborne ones. They also discussed how environmental health concerns, including water, are not within most clinicians' mindsets; they felt that there was no need to deal with something they do not see as broken.

Half of the group indicated that they had relationships with a S/LPHA, and that these links were easy, direct, and positive. However, they noted that in crises the nature of relationships changes. Routinely they see the S/LPHA and hospitals as good resources for information. Most indicated that their medical training did not emphasize relationships with S/LPHAs; some were not exposed to these linkages at all.

When asked about patients' concerns, clinicians said they talk with them generally about avoiding dehydration, but they put water in the perspective of their patients' vulnerabilities and more pressing health issues. Typically, water is not a priority issue with or for their patients; patients do not ask about water because "they take it for granted." When patients do ask, the clinicians have the impression that their issues are often influenced by recent mass media stories. Clinicians also said that they did not have environmental health training, so would need some orientation to address their patients' concerns.

From their viewpoint, clinicians identified several things that they felt facilitated collaborations: at the outset convening the directors of organizations, conducting drills that included clinicians, understanding each others' roles (especially who has the lead for which issues), developing trust over time and working together, tailored health education activities, addressing myths about water-related health issues, building broader contexts for water-related concerns, making sustainable commitments to each other, not placing blame, and relying on S/LPHAs to evaluate water-related health issues and forwarding information to clinicians as necessary. See Table D.3 in Appendix D for quotes representing key points made by clinicians.

Others. The academic and elected official representatives in this group emphasized the importance of personal relationships when addressing water-related concerns, building those relationships early, and engaging in joint problem solving. They noted that some relationships develop through events that *must* be addressed, that people discover they can work together, and then look for further opportunities to collaborate; not all partnerships occur through conscious, orderly effort.

Ongoing education about water issues was noted as essential; drinking water messages must be science-based and repeated for the public to develop their knowledge effectively. This group recommended that water issues be kept visible, upfront, and above all honest. The participants expressed concerns, however, that fears of litigation prevented officials from being as upfront as was in the public's best interests. A communication system needs to be in place in advance of adverse events; respected peers and spokespeople – particularly for subpopulations – need to be identified early as well. One way to involve subgroups is to include their representatives on committees that address water-related issues. In some locations, innovative partnerships, such as through local foundations or community groups, may be effective means to build relationships.

This group suggested that health agencies give talks to community groups, answer the public's questions, and write materials that could be broadcast faxed to communities. They also recommended that messages for communities be kept short, relevant, and to the point. The group emphasized that academicians can serve as resources for S/LPHAs and can conduct evaluations of programs for these agencies.

This group was clear that they trust government and utilities to do their jobs and thereby protect the public's health. When they discussed clinicians, these stakeholders felt that HCP education should be targeted where and when it was needed. Last, this group requested that utilities, PHAs, and clinicians consider academic extension services as resources, be available to the media, use teachable moments, and repeat public messages consistently.

What works from their viewpoint are seeking opportunities to work together, working on difficult problems together to discover common ground, having a key party who stimulates constructive interactions, developing cultural competencies, and ensuring that the needs of subpopulations (such as immune compromised individuals) are addressed. See Table D.4 in Appendix D for characteristic quotes made by other participants.

Region Specific Groups

The utility representative for each region was charged with identifying at least one health department employee and one clinician to participate in the workshop. Each region was given three additional slots to fill as they wished for this data sharing and gathering event. These additional slots were filled, for example, by additional utility or public health agency representatives, academicians, or elected officials.

Glendive, Montana. Glendive is a small town in rural, Eastern Montana. The water utility is part of the Department of Public Works; the health department is a County department based in the town. The health department serves a much larger geographic area than does the water system. There are eight clinicians who are affiliated with the local hospital in the center of town.

The participants from Glendive recognized that not enough officials are familiar with the area's drinking water system and what could happen to harm it. They also noted that the LPHA's communication plan has limitations and concluded that, as a community, they could improve their communications among utility, health department, and clinical personnel.

From their discussion, they decided that one way to improve their readiness would be to expand their practice scenarios beyond bioterrorism into more likely drinking water issues. Other ways would be to educate first responders, reshape information for dissemination in meetings, and improve the area's evacuation plan.

Lansing, Michigan. This capital city region in central Michigan encompasses town and county jurisdictions and urban and rural water systems, but the regional Ground Water Management Board serves as a center of jurisdictional information sharing and analysis for groundwater issues. The Ingham County health department is based in this town, and a major academic center, Michigan State University, is based in neighboring East Lansing. Although the public health department's jurisdiction is different from any single or the regional water jurisdiction, the participants viewed the regional communication of water issues as a very strong asset; it provides a hydrologic basis for their activities. This permits the utilities to convene and discuss concerns collectively.

Although this region has an emergency response plan and participants felt that they knew how it works, they decided that they needed a formal process to address the full range of issues that may occur. They recognized that they may be able to identify common issues using the Philadelphia Water Quality Committee model (see Appendix A), but they said they would need more information about this model before attempting to use it.

Other means to address common interests were also discussed. For example, the participants felt that the region would benefit from having a Memorandum of Understanding in place among the entities represented on a water quality committee. Additionally, health and clinical personnel requested a tour of the water utilities' facilities to increase their knowledge of utilities' operations and issues.

To address frequent questions from customers, utility personnel stated that they would like to have information from the county health agency about disease outbreaks, even if the outbreak was not related to water.

New York City, New York. This extensive metropolitan area encompasses five boroughs, some separated from others by rivers, but there is one health department and water purveyor (in the Department of Environmental Protection) that serves all five boroughs. This was the only region in this study that had geographic alignment for its public health and water services. There are thousands of clinicians, including a wide variety of specialists, who work in New York and serve its population, and patients from around the world.

New York City has a City Incident Management System (CIMS) that governs all agency activities in an emergency. Participants noted that, although the system is very good, continuous improvement is important. They also noted that the city has an established mechanism (i.e., broadcast fax system) for reaching the clinical community, but it probably does not reach all clinicians and should be reviewed for improvement. In particular, it is difficult for the LPHA to identify and reach clinicians who do not have hospital affiliations. Even when they can reach clinicians, the LPHA has to recognize that many clinicians, even those who acknowledge that water issues can be important, do not feel they have the time to deal with water-related concerns.

The participants stated that they need two things to improve health agency-clinician communications. First, they considered the development of annual brochures for clinicians to describe specific issues and list contact information (phone numbers and websites) for those interested in additional information. Second, they recommended the improvement of the New York City communication systems with clinicians, especially determining which clinicians they are missing and how best to reach them. The participants indicated that they would set up a meeting to further their ideas and develop an action plan to improve communications.

Redmond, Washington. This suburban town lies east of Seattle; its public health services are provided through the Seattle-King County Health Department. The public water services are administered through the town's Department of Public Works. There is one public and private hospital in or close to the town; many clinicians, including specialists, serve the town's clinical needs. This town has a large proportion of people who commute both into and out of this township. It was noted that, despite the town's modest size, the clinicians serve patients who come from a wide geographic area.

The participants from this area indicated that utility-LPHA and utility-federal/state agency communications need to be improved. Some concerns were raised that clinicians are not aware that many people are serviced by private wells and not public water systems. The group felt that clinicians often do not know what resources and materials exist to address water-related concerns.

Participants noted that physicians respond to vulnerabilities, particularly as they relate to their patients, so it may be useful to explore what problems clinicians perceive as relevant to their care-giving role. It was also seen as valuable to determine whether clinicians would like to have a role in addressing water issues.

The group recognized that above all the public and consumers want reassurance that their water is safe. They also noted that the public needs practical and relevant information that they can use readily.

Tucson, Arizona. Most of this arid town, and ten adjacent areas, receive their water from Tucson Water, based in Tucson. The public health department is organized on a county level, covering all of the areas served by Tucson Water. The town is the home of University of Arizona and has many clinicians, including a wide range of specialists.

This group focused on processes that they could use to improve their existing collaborations related to water issues. The LPHA was viewed as secondary to the utility in addressing these concerns. Tucson Water has taken the lead in building partnerships with communities and other stakeholders over the past decade.

The participants decided that they needed to establish a core group, with backups, to consider communication needs and mechanisms further. They indicated that they wanted to convene the core group, develop contingency plans, write a MOU, and identify any additional issues.

Recurring Themes. Water utilities indicated that their challenge is in figuring out what is abnormal when they get positive results for contaminants in water samples. They always report to the state and local health departments, and retest when the agencies see anything unusual. In addition, utilities say they must address the structure and level of communication and how to get the communications cycle to work to develop more effective three-way communications. They could do this by meeting on a regular basis and defining their functions by MOU or other structure, or improving their communication by drilling emergency scenarios with environmental health colleagues. Finally, utility officials suggested that there have been jurisdictional problems under emergency conditions.

Public health agency officials said that interactions with clinicians are ongoing issues in public health. Part of the problem with communications, they say, is in having adequate resources. Finally, they indicated that they need internal structure and contingencies in place so that the absence of critical players doesn't have serious impacts on the program. Clinicians pointed out they are generally missing from the discussion. Others need to frame water-related issues as vulnerabilities, so they will consider them as important issues. Other stakeholders said that vulnerable populations must be identified upfront, before problems occur. They also indicated that they need to look at what the public wants to know.

Lessons Learned

Several key themes emerged from these group discussions.

- All groups saw positive relationships – built slowly and consistently over time – as important to success in handling water-related health issues; they should not be expected to just happen during a crisis.
- Participants stressed that problems need be to be addressed upfront with a maximum degree of trust and credibility.
- Participants felt that generally the public has growing concerns about water issues.

- Utilities noted that their relationships with S/LPHAs were generally good, but varied depending on whether the PHA was regulatory or not. They expressed frustration that S/LPHA staff could not interpret utility data; S/LPHAs said they wanted to know how to read these reports.
- Utility and S/LPHA personnel view clinicians as lacking in the knowledge about water systems and issues to answer their patients' questions, which they feel is a crucial role of clinical staff.
- Some participants felt that a utility's purchasing of environmental health specialists' time for a few days a year may be well worth the investment to ensure clinician involvement for meeting both ongoing and emergency needs.
- Clinicians trust government and utilities to protect their patients' health, and expect utilities and S/LPHAs to reach out to them with information they need. They view S/LPHAs as credible vettors and communicators of water-health information.
- Both clinicians and utilities rely on LPHAs to serve as a bridge between them.
- The regional discussions produced suggestions for improvement in communications including: improving response plans, developing formal instruments and processes of communication, sharing data and information about their operations, revising written materials for each other's use, and establishing a core group to advance collaboration.

KEY POINTS

Utilities and health agencies realize that they have different interests, information and somewhat different obligations to the public. Neither utility personnel nor clinicians[*] indicated a need for closer direct relationships with each other. Although utilities and clinicians want S/LPHAs to serve as a bridge between them, in most cases S/LPHAs do not feel that they have the resources to fulfill these expectations. Further, S/LPHAs do not have complete lists of physicians and have limited or no lists for other clinicians whom they may need to reach about a water-related health issue. In some areas, S/LPHAs have not taken the initiative to develop contact lists and in other areas they have left the responsibility to clinicians to contact them to be placed on the contact list. Some S/LPHAs do not see it as their responsibility to identify all clinicians who might need to know about waterborne health issues.

Clinicians want utilities and PHAs to contact them, but utilities said they want clinicians to contact and inform them of their needs. Clinicians indicated that if they want to know about the water system, they want to talk with technical personnel who can speak their language; they do not want to be directed to someone who will waste their time. S/LPHA staff members reported that they contact utilities' PIOs as the gateway to the appropriate utility person to meet their needs. Typically, it is easier and more comfortable for clinicians to contact S/LPHAs, because they feel that they can contact high-level personnel who will understand and be responsive to their interests and needs.

As long as perceptual and functional communication barriers remain unaddressed, it is unlikely that utilities, S/LPHAs and clinicians will be able to initiate and sustain substantive relationships.

---

[*] In this case, most of the clinical participants were physicians.

The themes common across the literature review, and individual and group interviews were:

- Water-related risk communications among utilities, health agencies and clinicians are not ideal; there is room for improvement.
- Utility, S/LPHA and clinical personnel want more communication and collaboration. However, clinicians may not want or need to be engaged in water-related health risk collaborations as intensively as utilities and health agencies.
- Utility personnel and clinicians would like S/LPHAs to serve as the link between them, but these agencies are limited in their capacities to fulfill this expectation.
- Knowing specific persons to contact and the best way to reach them is crucial to effective communications among collaborators.
- There is no substitute for long-term, trust-based relationships among utility, S/LPHA, and clinical personnel. These require firm commitments and are facilitated by top managers setting the example. Such commitments could be explicitly conveyed in formal MOUs (e.g., NSWHD 2001).
- Clinical, utility and S/LPHA staff members would benefit from understanding more about each others' domains; e.g., roles, responsibilities, legal authorities, knowledge basis, etc.
- Utility, S/LPHA and clinical personnel all expressed judgments about each other and past experiences that could discourage the development of collaborations.
- Jointly preparing and implementing specific instruments (e.g., emergency response plans, MOUs and other tangible commitments) can facilitate establishing relationships and building long-term collaborations.

CHAPTER 3
EXAMINATION OF POTENTIAL COLLABORATIONS

INTRODUCTION

Literature reviews, surveys, interviews, and focus groups provided complementary methods for exploring the experiences and perspectives of professionals working in utilities, PHAs, and clinics as they addressed drinking water-related health issues. Some of these approaches yielded similar results, but using this variety of sources expanded the insights possible from any one method.

This chapter has two purposes. First, it highlights key concepts from business management literature relevant to making complex decisions and fostering collaborations. Second, the chapter presents the methods and results of the tabletop exercises. These findings were then used as the basis for finalizing tools to build collaborations, the Framework For Action, and the Cost-Benefit Scorecard (described in Chapter 4).

FOUNDATIONS FOR FIELDWORK

The tabletop exercises were focused on detailing each jurisdiction's interdisciplinary communication infrastructure, and revealing participants' goals and ideas about strengthening their capabilities to achieve more effective collaborations. The crucial bases for designing these exercises were (1) a thorough understanding of the evidence gathered in the earlier phases of the project, and (2) fundamental knowledge of the lessons learned in the business sector for managing dynamic and complex decision environments and problems. Not all lessons learned by other business sectors are pertinent to water utilities, but insights and key concepts were highlighted as we reviewed the literature. This chapter documents the most important concepts that aided our design and conduct of the regional fieldwork.

Evidence Base

As shown in Chapter 2, we identified the needs of each of the three sectors (utilities, PHAs, and clinicians) individually and collectively. Needs typically related to resources, knowledge, beliefs, perceived organizational barriers, and other factors. From the surveys and individual interviews we gained insights about each sector's needs, but the key opportunities for identifying and understanding the fundamental *reasons* for these needs came from the March 2004 workshop discussions (Chapter 2) and the Winter 2004-2005 tabletop exercises (in this chapter) conducted in each of the five participating regions.

To obtain information about each area's progress toward collaborative relationships and design the tabletop exercises, the Principal Investigator (PI) telephoned each of the utility leads four months after the workshop. In these conversations she listened for: (1) actions already taken to build collaboration, (2) concerns about and barriers to achieving collaboration, and (3) the major drinking water issues confronting each utility. She provided each utility lead with general information about the tabletop exercises and discussed plans to conduct them.

Literature Reviews

To develop the basis for this portion of the project, we reviewed peer-reviewed publications in the fields of risk communication, stakeholders, and business management. This multidisciplinary set of knowledge allowed us to select the most relevant bases for water-health risk communication collaborations involving both public and private organizations.

Organizational Approaches to Risk Communication

First, we examined how various organizations approach risk communication; what types of collaborations they have, how they are structured and what influence that has on risk communications, etc. What we found is that, although risks are often discussed in terms of individual perceptions and actions, organizations have important roles in shaping the contexts for risks. A central task of organizations is "the estimation and management of two types of risk: risks in the environment that challenge operations, and risks associated with the products and operations. Due to their ubiquity and power, organizations have the prerogative to frame and define public issues, including risk issues" (Jaeger et al. 2001). Though much of the literature about organizational roles is targeted to corporate interests, the principles are largely applicable to utilities, public health agencies, and other public organizations.

Complex risks – such as drinking water-related health risks – are multi-dimensional and need to be framed comprehensively. By working with many types of partners (including the public) to improve risk-related decision-making, organizations can be better positioned to address the multiple dimensions of risk, and can produce decisions that are more relevant to all parties involved (Slovic 2000).

One pervasive component of risk management is risk communication (CSA 1997), but it is understood and defined in many ways. The most widely accepted definition of risk communication was published by the National Research Council (1989):

> "An interactive process of exchange of information and opinion among individuals, groups, and institutions; often involves multiple messages about the nature of risk or expressing concerns, opinions, or reactions to risk messages or to legal and institutional arrangements for risk management."*

This definition may seem overwhelming when organizations first approach a collaborative decision-making process. But the important thing is to remember that the goal of effective risk communication is to inform decision-making. However, the complexities of risk issues in modern democratic societies, especially those with Internet capabilities, require more complex risk communication strategies that are more advanced than either "tell" (one-way) or even dialogue (two-way) approaches (NRC 1996b, Jaeger et al. 2001).

Further, in the United States there is a cultural emphasis on collective decision making for major issues involving many entities; this norm demands that attention be focused on assuring procedural integrity through rigorous application of mutually agreed criteria. By sharing responsibility and creating a structure and shared goals, organizations can engage in effective discourse,

---

* The necessity to address risks on many different levels has been extensively described, but was perhaps best presented by Slovic (2000).

Table 3.1
Key traits of stakeholders

Trait	Definition
Power	The extent to which a party has or can gain coercive, utilitarian or normative means to impose its will in the relationship.
Legitimacy	A generalized perception or assumption that the actions of an entity are desirable, proper, or appropriate within some socially constructed system of norms, values, beliefs, and definitions.
Urgency	The degree to which stakeholder claims call for immediate attention, under two conditions: (1) when the claim or relationship is time-sensitive and (2) when that relationship or claim is important or critical to the stakeholder. [NOTE: These two criteria must be from the stakeholder's perspective, not the lead organization's.]

Source: Adapted from Mitchell et al. 1997.

and thus be better prepared to serve the public's good (Jaeger et al. 2001, Magee 2003). Drinking water utilities, public health officials, and clinicians could utilize this or other types of deliberative processes to create stronger collaborations (e.g., NRC 1996b).

Stakeholders

Managing stakeholder relationships has become a key area of business literature. Lim, Ahn, and Lee (2005) noted that there are four strategic phases that are especially useful in complex, multi-stakeholder environments: stakeholder analysis, strategy retrieval (or reassessment), strategy revision, and implementation. Lead organizations are advised to monitor stakeholders' interests to determine whether a group's significance to the organization is increasing, decreasing, or staying the same. Interacting with stakeholders requires a dynamic, timely strategy appropriately tailored to each group's importance under prevailing conditions.

Insights about how to identify and prioritize key partners or stakeholders have been detailed by Mitchell, Agle, and Wood (1997). Their comprehensive review of organizational management literature provides a structure for recognizing, characterizing, and understanding the interests of stakeholders as a basis for effective organizational strategies. The authors use the long-standing definition of stakeholder (Freeman 1984), "any group or individual who can affect or is affected by the achievement of the organization's objectives."

Eight groups of stakeholders have been defined and their significance to organizations recognized on the basis of three traits: power, legitimacy, and urgency (Table 3.1 and Figure 3.1 derived from Mitchel, Agle, and Wood (1997)). Power is defined based on coercive, utilitarian, and normative characteristics; e.g., whether the stakeholder possesses the capacity to force its will on others through its use of physical force, material resources, or symbolism (Etzioni 1964). Legitimacy is defined as socially accepted and expected structures and behaviors that are generally viewed as appropriate and desirable, within socially derived norms, beliefs, and definitions (Suchman 1995). Urgency is defined as deriving from stakeholders' awareness that their claims are critical and time-sensitive; e.g., their concerns require immediate attention (Mitchel, Agle, and Wood 1997).

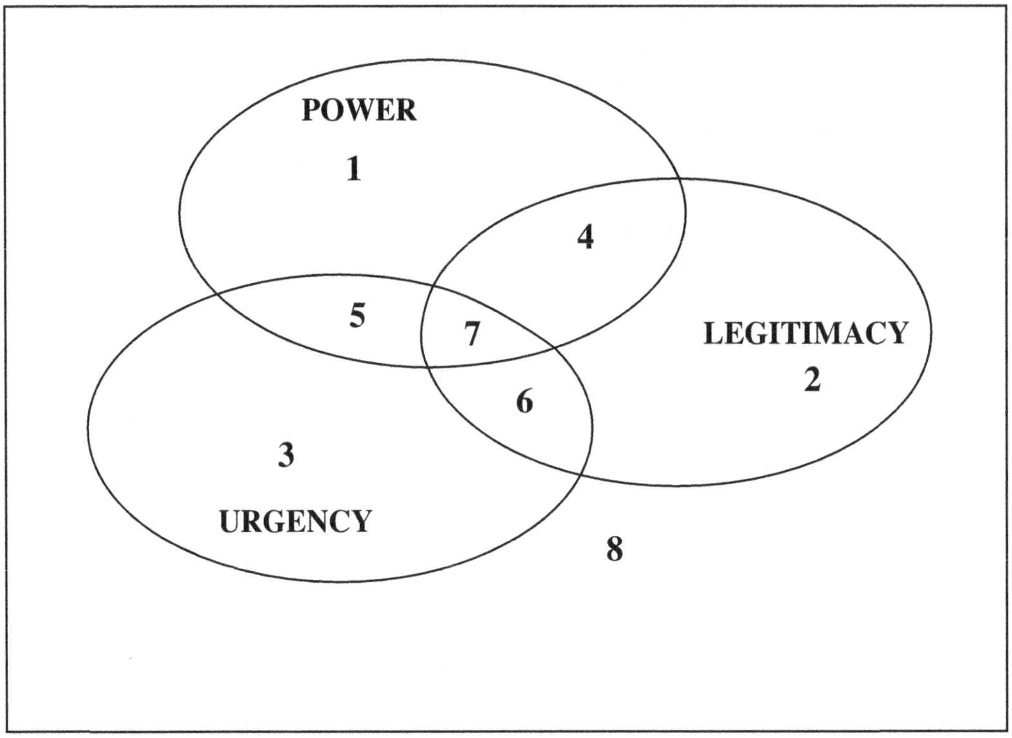

1: Dormant	3: Demanding	5: Dangerous	7: Definitive
2: Discretionary	4: Dominant	6: Dependent	8: Non-stakeholders

Source: Adapted from Mitchell, Agle, and Wood 1997

Figure 3.1 Stakeholder typology

Driscoll and Starik (2004) criticized this framework and added an additional characteristic – proximity – but the original concepts and approach laid out by Mitchell, Agle, and Wood (1997) have been applied in dozens of cases and have provided valuable insights for advancing organizational collaboration strategies.

It is important to note that organizations may have incorrect judgments of stakeholders (Argyris 2001, Hammond, Keeney, and Raiffa 2001) that can become consequential in designing decision-making and risk communication strategies. However, these errors can be detected through rigorous data gathering, information sharing among collaboration partners, checking one's assumptions, and active, ongoing surveillance for changes in societal and other contexts (Etzioni 2001).

Both known and potential stakeholder relationships must be considered because stakeholder traits can change over time, and may do so especially when time-sensitive concerns arise. Societal and/or political contexts (e.g., an election year) may have significant bearing on whether stakeholders change their alliances and move to a different level of relevance to the organization. Also, stakeholders may adopt the traits of other stakeholders by forming alliances; e.g., concerned communities may increase their urgency by linking with advocacy groups, or their power by getting the support of elected officials. Similarly, clinicians may increase their power by working through PHAs, or their urgency by aligning with advocacy groups. Additionally, stakeholders typically seek access to organizations through four mechanisms: linking with internal subunits, utilizing legislated participation requirements, accessing legislated information, and undertaking activism initiatives (O'Connell et al. 2005).

Decision Making

The recent, rapid growth of information does not necessarily lead to growth in knowledge. Organizational managers may want to deny or avoid dealing with this reality, obsess on gathering evidence rather than taking action, or ignore emerging political or stakeholder concerns that may ultimately force action (Etzioni 2001). However, because of the complexities of new knowledge and the interdisciplinary nature of many problems (including drinking water-related health risk communications), current conditions require greater interaction and collaboration than ever before. Without comprehensive knowledge of legal, social, and other contexts, it is difficult – if not impossible – to develop and implement effective communication strategies and collaborative partnerships (NRC 1996b).*

Creating collaborations in any environment is a challenging, iterative process. Yet experience and research have revealed certain commonalities among effective decision processes. For example, Drucker (2001) noted that there are six steps in effective processes:

- Classifying the problem
- Defining the problem
- Specifying the answer to the problem
- Deciding what is "right," rather than what is acceptable
- Building into the decision the action to carry it out
- Testing the validity and effectiveness of the decision against the course of events

He also stated that the key to sustaining relationships and making effective decisions is checking over and over again to see whether anything unexpected is occurring. He exemplified this by referring to clinicians' approach to diagnosis and treatment. Doctors make decisions incrementally (not committing all resources in one step but seeking out the latest clinical guidance), and check to see whether their decisions are causing the expected results. If not, clinicians reassess the situation and make adjustments, always keeping in mind the ultimate, desired outcome. Drucker (2001) emphasized that effective managers make decisions through a systematic process involving clearly defined elements in a series of distinct steps.

Etzioni (2001) recommended a "mixed scanning" method of organizational decision-making; i.e., broad, basic choices about the organization's policy and direction, along with step-wise decisions that prepare the way for the implementation and realization of current and future judgments. Like Drucker (2001), he stated that this is the approach physicians have used successfully for a long time. However, for this method to work on an organizational scale, managers' behavior must match their attitudes (Argyris 2001). For example, expressions of distrust tend to destroy the potential for collaboration, while explicit opportunities for open dialogue and feedback foster a cooperative environment.†

---

* See Chapter 6 of NRC 1996b.

† Hammond, Keeney, and Raiffa (2001) have provided more details about how managers may inadvertently contribute to poor decision processes.

Decision Analysis

One phase of informed decision making processes involves evaluating and interpreting evidence. Decision analysis is a systematic, stepwise approach for identifying and assessing the relative value of different decision options based on such evidence. The method uses formal tools; focuses participants on defining the problem, and stating objectives and measures; and involves facilitated processes to produce informed judgments for addressing the problem. Conducting decision analysis processes promotes creativity, provides new insights, and fosters shared commitments to a final "best" option. The quality of the process depends on whether the analysis is based on sound theory and evidence, is applied systematically, and whether it captures key elements from the full context of the decision (Brun 1994).

Structured decision analysis methods are especially useful when decisions are complex; e.g., when problems have many dimensions, information is uncertain, and stakeholders have a wide range of values and opinions related to the problem (von Winterfeldt 1992, Keeney and Raiffa 1993, Pettiti 2000). These characteristics are typical of drinking water-related health risk communication decisions.

Although the decision analysis process may be quantitative, the components of the method can be adapted to more qualitative tools. The resources available to support the process and the type of decision/s to be made are important factors to consider when determining whether the process should be more or less quantitative. Mobley et al. (2005) have stated: "Regardless of size, a water utility needs a framework and a process for strategic communication planning."

Scenario Planning

We reviewed scenario planning literature and related examples to ensure successful design of the tabletop exercises (e.g., Ogilvy 2002, Lindgren and Bandhold 2003). A scenario is a "hypothetical sequence of events constructed for the purpose of focusing attention on causal processes and decision-points" (Kahn and Wiener 1967 cited in Cooke 1991). Scenarios provide a mechanism to systematically and strategically reveal a community's shared values, fears, and aspirations for intermediate and long-range problems. The sequence of steps is the same as for decision-making processes, but the quality of the process depends on (1) developing realistic and relevant situations to work through, and (2) gathering people who view the situation in diverse and comprehensive ways, and who share key values and attributes; e.g., trust, confidence, generosity, patience, respect, creativity, systems thinking, and openness to intuition and discovery. As participants challenge their own thinking, they build skills to question "group-think" traps. When values and hopes become shared, they provide the basis for a "tipping point" that leads to new commitments, collaboration, and vision of the future.

Effective scenario exercises result in new insights, improved planning and preparedness, identification of gaps and needs, a broader range of options than any one group of individuals would have recognized, memorable experiences, shared learning, recognition of counter-productive assumptions, stronger interpersonal relationships, as well as new confidence and skills for addressing complex problems. Successful exercises require frameworks, tools, facts, processes, linkages, and facilitation. In conducting the exercises, our goal was not only to facilitate collaboration in the participating regions but also to learn lessons for improving our draft Framework.

TABLETOP EXERCISES

The workshop discussions confirmed that among the five participating areas, there were important differences in organizational structures, legal authorities, interests, commitments, and shared histories with collaboration. While "tabletop exercises" are typically designed to test a specific protocol for responding to a range of conditions (e.g., Emergency Response Plans for a variety of emergencies), the workshop results indicated that no single framework would be suitable for conditions in all five areas. Additionally, the follow up phone calls indicated that limited progress toward collaboration had been made. As a result, a needs assessment and planning tool, rather than a response protocol, was more appropriate and useful for the participants.

Methods

The exercises were conducted in the five participating regions: Glendive, Montana; Lansing, Michigan; New York City, New York; Redmond, Washington; and Tucson, Arizona. Personnel from water utilities, state and local health departments, clinics and hospitals were included in all five sites. Representatives of academic centers, poison control centers, emergency response offices, Mayors' offices, and other local governmental agencies participated in some but not all regions.

Each field exercise included two scenarios, evaluation of the draft Framework For Action, and evaluations of the entire day's program. The agenda typically required six to seven hours to complete.[*]

Draft Framework for Action

The draft Framework was designed to guide the participants through a structured discussion of the following topics:

Assessing their Status Quo

- Public risk communication materials; e.g., types, purposes, accessibility, date of last update, etc.
- Means of communicating with each other and other stakeholders, including types and frequency of contact
- Existing legal authorities, regulations and mechanisms related to water-health issues (e.g., bases and methods to call for and end a Boil Water Alert, provide public information and education, etc.)
- Subpopulations that may be at increased risk from water contamination

Determining Bases for Progress

- Principles and goals desired in a three-way collaboration
- Timeframes in which the goals could be reached
- Potential partners or stakeholders who could contribute to a collaboration

---

[*] All materials and procedures were reviewed and approved by the GWU IRB (IRB #U070222).

Participants were directed to focus on one topic at a time, to complete it to the best of their ability in about ten minutes, and then to orally exchange their answers and views.

By stepping through the draft Framework one section at a time, the facilitator was able to determine at each step which section would be the most appropriate step for the participants to address next. As a result, all sections were not used in all five areas, but all sections were used in at least two of the five areas. This approach allowed for flexibility in meeting each area's needs, while generating data from more than one area for each section of the Framework.

Scenarios

Each site worked through two different hypothetical drinking water-related health scenarios. The five scenarios included some of the common challenges mentioned by utility representatives in the follow up telephone calls. Each scenario was tested by two utilities and their partners, but no combination of scenarios was repeated (Table 3.2). Using each scenario twice was necessary to examine the robustness of the Framework and decision-making issues across service types, geographic areas, and communication situations. The scenarios provided to each utility were different in nature; e.g., one microbial and one chemical situation, or one acute and one chronic health hazard. Each scenario unfolded in a step-wise manner; no group received all of the information about the scenario at the beginning. New elements of the scenario were distributed as the group's discussion became minimally productive.

Scenario 1 was set around a town's annual fall festival. Several days after the festival, a number of community members fell ill. The health department indicated that the illness was unlikely to be the flu, so people began to think about other culprits of the outbreak, including possible microbial contamination of the water supply. Participants were asked to consider their roles and responsibilities in the scenario, how they would proceed in light of limited information, and who they might contact at various stages of the event.

Scenario 2 involved a company's proposal to the city council to build a large-scale golf course. Opponents of the proposal were concerned about chemical runoff from the facility, including nitrates. Participants considered how their organization would respond to questions about community members' health concerns, how doctors would get information they might need to treat patients or answer questions, and how their organization might respond to information (of unknown validity) other experts were giving to the public.

In Scenario 3, participants were asked to respond to national news media reports about long-term exposure to low levels of pharmaceuticals and personal care products in drinking water. They were asked where they went for information about the issue if they did not already have it, and what their concerns would be. Participants reacted to hypothetical reports of anti-depressants in nearby drinking water, and described what they would say at a city council hearing on the issue.

Scenario 4 examined an accident of a truck carrying hazardous materials. Participants were asked how they would proceed immediately after the accident – what their responsibility would be, what information they would need and how they would get it, and who they would need to work with. They also discussed appropriate roles and responsibilities for long-term issues related to the event.

In Scenario 5, participants were asked to react to press coverage and public concern about possible health effects from exposure to water disinfection byproducts. They were asked how they would respond to community groups asking utilities to "do the right thing," how they would answer doctors' questions on appropriate patient care, where they would go to get additional

Table 3.2
Distribution of scenarios by utility

Scenario	Utility A	B	C	D	E
1. Acute microbial contamination and related illness		x			x
2. Emerging chemical contamination & potential health outcomes	x		x		
3. Long-term chemical contamination and adverse health effects	x			x	
4. Acute health effects from a chemical contaminant			x		x
5. Potential chemical contamination and health effects		x		x	

information, and what they would recommend to their supervisors for a longer-term response to the issue.

Format. The exercises were designed to provide participants with sequential sets of information about the scenario and then guide them through a group of questions after each set of new information. The scenarios involved periods from seven days to one year. The PI facilitated the sessions, and one of two research assistants recorded on tape and in writing the discussions of each exercise. The participants were given both written and visual materials, and offered technical support at the end of each exercise. The investigators received participants' feedback on each exercise, the draft Framework For Action, and the entire tabletop program. At the end of each day, participants completed written evaluation forms to provide the researchers with feedback about the activity.

Typically the participants used the draft Framework For Action between their two scenarios. At the first site, the Framework was used at the end of the day, but the participants had low energy to address the issues at that time. Therefore, the remaining four sites worked on the Framework between their two scenarios. As with the scenarios, the PI facilitated and the research assistant recorded the groups' discussions as nearly to verbatim as possible.

After each site visit, the assistant transcribed her notes, using the tape to assure accuracy. The PI reviewed the transcripts for clarity, compared them to her own field notes, and conducted a systematic qualitative analysis to uncover the major themes from the discussions. These themes were examined to assess whether they occurred universally, or in specific types of jurisdictions or geographic locations. The results were used as the bases for developing the final Framework and Cost-Benefit Scorecard presented in this report (see Chapter 4).

Participants. Seven to ten people participated at each of the five utility sites. Men and women were present for all exercises, but only two sites had minority (Hispanic, Native American, or Asian) participants. The employers of the individuals who attended the exercises were water utilities, PHAs, medical facilities, other local organizations (e.g., fire department, poison control center), and universities. Utility staff roles included engineer, water quality manager, drinking water analyst, manager of water operations, and emergency response planner. PHA participants' roles included public information officer, emergency or hazardous material (hazmat) responder, director of environmental health, director of constituent relations, water engineer, drinking water-health expert, toxicologist, epidemiologist, sanitarian, county health official, and medical officer. Three of the five areas had up to two participating clinicians who were not employed by the LPHA; one area had two physicians who were each present for different half-days and another area was able to retain one clinician for the morning session only. Participating clinicians' areas of specialty included internal medicine, sports medicine, pediatrics, and HIV/AIDS. Other participants included a public works director, and university faculty and graduate students in toxicology, ground water, or health risk communication.

Findings

In all cases, the participants engaged in thoughtful conversations throughout the scenarios and Framework exercises. Although some individuals wanted more detail in the scenarios, the facilitator emphasized that the scenarios did not need to be "solved" as would be the case in more traditional tabletop exercises. The important activity was to respond to the questions posed (e.g., about actions that would be taken, who would be doing what, and what concerns would be confronted in responding to the scenario conditions). The discussions that ensued in answering these questions provided rich information about regions' capacities to respond to different types of water-related health risks, participants' knowledge and assumptions about each other's responsibilities, and gaps in preparedness. Themes identified from the scenarios and Framework were very similar. Thus, the primary findings from all components of the regional exercises are presented together in the following section.

Legal Authorities and Mandates

When assessing the participants' knowledge of the authorities and mandates that underpin each organization's activities related to drinking water health risk issues, it was apparent that most people had limited awareness and knowledge of the legal bases for their programs. One health official noted that the state public health laws were outdated for addressing these water-health risk issues.

Roles and Responsibilities

Utility, PHA, clinical, public information, and emergency response personnel were unclear about each others' roles and responsibilities in most of the scenarios. During the exercises, clinicians asked more questions about the other participants and organizations than did other personnel, but there were important gaps in knowledge among all categories of participants. Utility and PHA professionals had to ask questions about each others' policies and procedures to complete most of the exercises. Realizations about their interrelationships became apparent in the discussions. For example, one participant said:

> "I've realized that we don't always agree, but we do have to talk."

Another expressed his concerns about the status quo saying that the lack of alignment in utility and PHA policies and procedures is:

> "A gap that may come back to haunt us."

In some cases, participants made assumptions about each others' operations that led to misunderstandings; some of these assumptions could block effective responses during actual events. For example, a PIO acknowledged that she did not realize the problems PHAs have in communicating with clinicians. She said she assumed that the health department contacted doctors and added:

> "It's a governing myth....I never realized that not everyone would be contacted."

This assumption was found in all five areas; participants in all areas appeared to be discussing this issue with each other for the first time.

Another observation was that participants were not always aware of what they did not know that had consequential impacts on their relationships. Most notably, although many participants were familiar with the term "risk communication," many expressed frustrations during the exercises that resulted from their limited understanding of risk communication strategies and options. Many times the speakers indicated that they talk *to* others, not appreciating that an interactive or dialogue approach may work better. For example, one utility representative said:

> "Recently we linked to doctors. We tried to talk *to* them about *cryptosporidium*, waterborne diseases ..." [emphasis added]

Several people expressed commonly held views about communicating with the public:

> "We have to describe the problem *to* the receivers, what problems you face, what information you have." ["You" here refers to the speaker's organization.] [emphasis added]

> "It's a no-win situation. We don't know what to say."

> "Any emerging issue can be a landmine!"

Numerous times it was clear that participants had important boundaries on their conceptualizations of risk communication.

A particular issue that arose in four of the five participating regions related to PIO functions. One region did not have clear procedures about press activities, including how press statements would be approved for use or who would provide these statements. In another area which did not have a defined PIO, a participant flatly stated:

> "Our people are professionals [handling the press] ... We like to address things quickly – we've never had a problem."

Another region noted that their utility and health agency PIOs had a history of working closely, and had experienced increasing success. Joint press releases were often mentioned in the tabletop exercises. Most utilities viewed PHA staff and clinicians as sources who were more acceptable to the public for health information.

Issues arose about reaching subpopulations that may need different information in different ways than the general population. One region had made considerable progress in building community relationships through outreach and participation in community events. Two regions noted that they had made efforts, beyond translating written materials, to communicate with populations that did not use English as their primary language. Subpopulations recognized in the scenarios as potentially sensitive or special needs groups included dialysis patients, persons living with HIV/AIDS, patients receiving chemotherapy, transplant patients, elderly and very young individuals, as well as people who speak English as a second language. A utility staff member commented that:

> "We found out that ACT UP [an HIV/AIDS patient advocacy group] didn't support filtration but supported watershed protection; that's not what we expected."

No regions were clear about how best to reach most special needs groups, including patients with prior illnesses. Reaching out and understanding subpopulations' issues remain major needs and significant challenges in all areas.

Another striking gap in regional roles was that no organization seemed to have the responsibility for pretesting communication materials or assessing the effectiveness of communication methods including CCRs, translated materials, broadcast faxes to clinicians and medical facilities, etc. Although most areas had provided information to physicians about water-related health issues, none had evaluated either the materials used or the outreach conducted. Many participants appeared to have never thought of either pretesting materials or determining whether their communication efforts were having the intended impacts. While pre-testing and evaluation are often difficult, it was not clear that participants felt that this merited the same level of effort that they expressed about reaching special needs populations.

Relationships

Some regions had longer traditions of working together than others; some participants were very familiar with others, while some had not met any or most of the other participants. All areas indicated that they had had largely positive utility-health agency coordination, but none mentioned inclusion of clinicians in any of the historical examples described.

One issue that arose in one location was the short terms that people seemed to be in any one position, making it difficult to establish and maintain relationships necessary for substantive collaborations. One person noted that this level of mobility made it difficult for the region to serve the public's interests as the public expected. Another participant stated that he was gratified to find that many people who changed jobs actually went to similar positions within the region. Therefore, they were not entirely removed from potential partnerships for communication; in fact, his view was that their job changes offered opportunities to link with additional nearby jurisdictions with which he had not yet built relations.

It was striking that in all areas, participants had a narrow conception of the types of individuals included in the term "clinicians;" this was partly reflected in the fact that the only clinicians present in the tabletop exercises were physicians. Participants typically interpreted "clinicians" to mean physicians only, although in the exercises participants mentioned many other clinicians. Specialty physicians cited were oncologists, obstetricians/gynecologists, pediatricians, "kidney doctors," and urologists. Other clinicians and related personnel discussed during the scenarios included nurses (public health, community, and school nurses), veterinarians, pharmacists, hospital infection control officers, home health staff, toxicologists, Health Maintenance

Organization staff, hospital staff, laboratorians, nurse hotline subscribers, emergency room staff, poison control staff, and dentists.

Further, many of the participants were unclear about who could be considered "partners" for water-related health concerns. During the exercises, professionals cited as being needed to implement the desired response activities included police, academic clinicians and experts, private laboratorians, public and private school personnel, day care center staff, nursing home staff, and wastewater plant operators.

There was considerable discussion among the participants in two areas about their dissatisfaction with the fire and/or police personnel, in their responsibilities for emergency operations. They indicated that many times emergency operations debriefing sessions were not held at all, so that lessons that needed to be learned across all responsible agencies were not identified. When sessions were held, participants felt that the "police model of blame and finger pointing" was counterproductive for transforming problems experienced during real and drilled emergency scenarios into lasting lessons.

Utility and LPHA participants said that their organizations conducted internal debriefings, but that important opportunities to serve the public better were being lost by not having more effective interagency evaluations. Suggestions were made that a neutral facilitator, perhaps an outside expert, should run debriefings to remove the climate of blame and to help agency personnel move toward more constructive analyses and decision-making.

Communications With Each Other

Themes found in earlier phases of the project were also found here. For example, all five areas noted that the contact lists they had for physicians were incomplete, ranging from 1% missing to no updated list available. One area relied on physicians self-identifying for inclusion on the contact list. Other areas viewed the compilation and updating of the list as a LPHA responsibility, but LPHA representatives said they have limited resources to compile comprehensive lists or to keep them current. In two regions, suggestions were made that state licensing databases may be useful in keeping lists updated. Two areas noted that the lists should not be overused; e.g., information for physicians should be targeted selectively and for "big issues" only.

Most of the participants were unclear about how broadcast faxes were distributed once received by hospital and clinical personnel. Most believed that PHAs directed the faxes to the hospital infection control officer or the director of a group practice, but no one knew how effective recipients were in distributing the faxes to the clinicians and other personnel who needed to know about them. One concern was raised that an infection control officer might not know to whom a notice about chemical contaminants in water should be given. None of the five areas had conducted an evaluation to determine the effectiveness of the fax method in reaching physicians and other clinical personnel.

Additionally, S/LPHA and clinical personnel were not always familiar with terms commonly used by water utility personnel. A term as familiar as "CCR" in the water industry was not known to many of the public health and clinical participants, nor did they know that CCRs are legally required to be distributed annually. Water system operations and terms were unfamiliar to most of the non-utility participants, except in one area where the utility had hosted tours of their facility for health professionals. In this case, LPHA staff members commented on how interesting and useful the tour was in expanding their knowledge and appreciation for utility concerns and challenges.

While the communications during most tabletop exercises were constructively intense, a few times the stress from past experiences was released in statements that indicated limited respect or understanding of another participant's views, responsibilities, or professional abilities. Some of these tensions were quickly addressed and resolved by the participants, but in a few cases tensions remained that could compromise the region's professional climate and development of effective collaborations.

Several sets of participants noted that it was a lot easier to deal with the acute scenarios than the chronic or future water-related health concerns. Although one area had an individual who felt that the emergency response approach could be used for both types of concerns, that view was not shared by others present.

Stakeholders

Mitchell, Agle, and Wood (1997) typology (presented earlier in this chapter) provides a method for organizing the lead organizations' views of various stakeholders in their regions.[*] The investigators analyzed the transcribed tabletop discussions to identify the stakeholders mentioned in the exercises and what roles and responsibilities they were expected to have under routine and emergency scenarios. The authors then used the results from this analysis to place the stakeholders mentioned by the utilities (in this case, utilities were the lead organizations) into Mitchell et al.'s stakeholder categories (Table 3.3). As revealed in the scenario exercises, stakeholders were listed twice in the table (in the two right-hand columns); first, according to their characteristics described under routine conditions and, second, according to their anticipated roles.[†]

Further, mapping the stakeholder results provides an example of how Mitchell et al.'s typologies could be used to aid decision-makers in determining how and *why* stakeholders change salience to organizations (in this case, utilities), and provides insights into how to prepare for working with stakeholders under different conditions. For example, in Table 3.3 it is apparent that LPHAs and clinicians do not fall into the same categories (e.g., under routine conditions and in many scenarios, clinicians are "discretionary" while LPHAs are "dominating" stakeholders); in fact, clinicians typically derive greater salience to utilities through their relationships with LPHAs. There was no evidence in the scenarios that suggested utilities and clinicians needed to have a direct relationship for addressing acute or long-term water-related health issues. Scenario participants said that as long as clinicians are confident that LPHAs appropriately attend to their interests, it is unlikely that direct utility-clinician relationships will be essential. Utility representatives noted that when another organization is the lead for a crisis or routine issue, the utility's role will change and therefore may affect their relationships with LPHAs and clinicians.

---

[*] This analytic structure was not discussed in the workshop or tabletop exercises.

[†] The categorizations in Table 3.3 revealed how utilities viewed their stakeholders, not how LPHA, clinical, or other stakeholders see their own roles in routine or other conditions. The same types of listings could be created from the evidence for the other participants, but would likely produce somewhat different results. Comparing the listings for utilities, LPHAs, and clinicians would reveal similarities and differences in perspectives about stakeholders. The alignments and misalignments between collaborators' groupings would indicate opportunities and challenges respectively for utilities and their partners before and during non-routine conditions.

Table 3.3
Utility perspectives on stakeholders under routine conditions and in tabletop scenarios

Category*	Trait/s	Routine conditions		Tabletop scenarios	
1 = Dormant	Power	Board of Health		Board of Health	
2 = Discretionary	Legitimacy	Government agencies (typically not *LPHAs*) Hospitals, clinics Poison Control Centers Professional associations Voluntary organizations General public Community groups Academic centers	Fire, police Press Lawyers Industry **Clinicians** Consumers Schools Experts	Other government agencies (other local, state, USEPA, CDC) Hospitals, other medical entities Poison Control Centers Professional associations Voluntary organizations Community groups Academic centers Parent-teacher associations	In some: Press Lawyers Industry **Clinicians** Consumers Advocacy groups Schools (private) Experts General public
3 = Demanding	Urgency	"Worried well" Advocacy groups		"Worried well" Advocacy groups	"Disgruntled" public
4 = Dominant	Power and legitimacy	Utility Board of Directors **LPHAs** Elected officials		Other government agencies (e.g., regional EPA) In some: **LPHA*s***, Voluntary groups, Elected officials, Press	One scenario: Police Where directly involved in the scenario: Industry
5 = Dangerous	Power and urgency	In some regions: Press		Advocacy groups In some: Press	
6 = Dependent	Legitimacy and urgency	Subpopulations		At risk subpopulations (e.g., children, elderly, ill, etc.) Affected community groups	Affected schools **Affected area clinicians**
7 = Definitive	Power, legitimacy, and urgency	[no mention]		Emergency Response coordinator In some: **LPHAs** Utility Board of Directors Elected officials	Fire, police Press Public Press
8 = Non-stakeholders	None	[no mention]		[no mention]	

* Mitchell, Agle, and Wood 1997

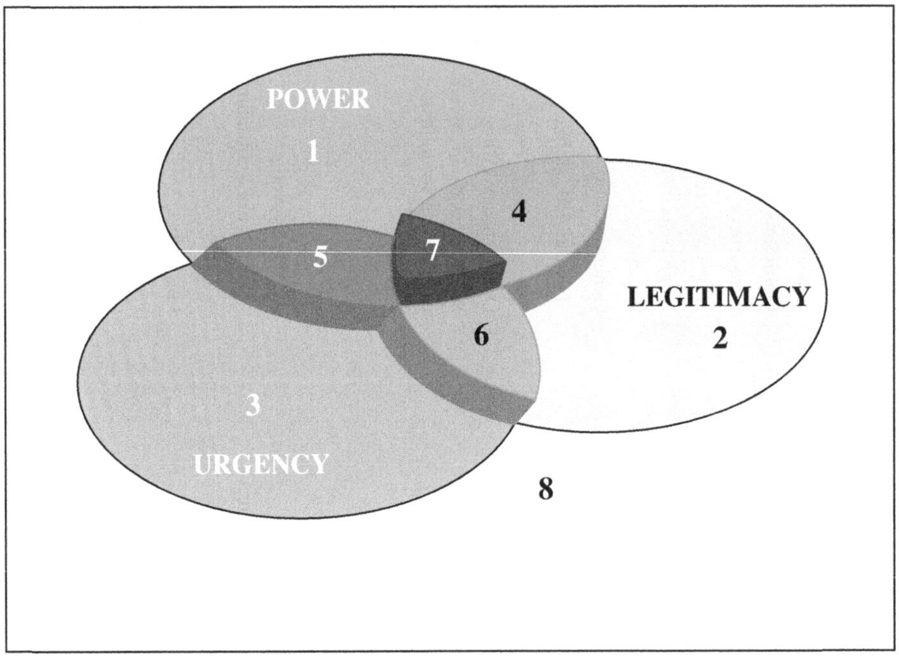

1: Dormant	3: Demanding	5: Dangerous	7: Definitive
2: Discretionary	4: Dominant	6: Dependent	8: Non-stakeholders

Figure 3.2 Recommended level of lead organizations attention to stakeholders

Last, this example demonstrates how stakeholders can be evaluated and sorted into groups that clarify the level of attention lead organizations must give to each stakeholder (Mitchell, Agle, and Wood 1997). In Figure 3.2, we have shaded the framework to indicate from low (light gray) to high (dark gray) according to the Mitchell, Agle, and Wood (1997) theory. They state that entities with Power and/or Urgency characteristics *must* receive attention. All others receive attention according to the strength of their traits, given that each trait is not an "either/or" so much as a gradient along a spectrum. In resource-limited or less complex conditions, using the framework on an "either/or" basis may be sufficient. The more complex the setting, the more detailed the evaluation of traits should be.

While some insights about stakeholders' interests can be gained from informal processes of gathering and interpreting evidence (e.g., talking one-on-one, attending public events, meeting with community leaders, observing Advisory Panel meetings, etc.), the best strategies for developing comprehensive knowledge about stakeholders and the level of attention required are based on rigorous, scientific methods, not guesses, about others' issues and contexts (e.g., Morgan et al. 2002). It is crucial to understand not only *what* people think and believe, but also *why* they think and believe as they do. Guessing what people think and believe is insufficient for developing appropriate collaboration strategies, especially for complex issues and heterogeneous populations. Guesses lead to incorrect assumptions and actions that may be seen by others as harmful. Such observations erode trust, the cornerstone of a successful collaboration, and may remove or severely delay opportunities to rebuild trust (Parkin 2005).*

Participants in this project repeatedly noted that they made assumptions about each others' roles, and thereby wound up with problems operationally. It was only when they identified effective means to reveal assumptions, understand their origins, and correct their false assumptions that they found collaborative efforts became feasible.

Principles and Goals

Participants were asked to consider what they felt the key principles and goals should be for a potential collaboration among utilities, LPHAs, and clinicians. Principles typically stated were: honesty, trust, respect, prevention (not reaction), no "blindsiding" of each other, and relevance to organizational needs and priorities. Openness was also noted, but in one site this principle led to considerable debate about how open they could really be with each other. They said primarily they have to be loyal to their agencies, and did not all feel that "off-the-record" conversations were appropriate. A quote that highlights a crucial issue from this debate is:

> "You don't gain trust all at once. You can be more open after you gain trust. You can't put things out right away without knowing a person. You need to know people's motives and behaviors first."

Baseline values to support collaboration that were mentioned less often by the participants included: awareness of each other's roles and responsibilities, timely responsiveness, accuracy, routine communications, clear goals and purposes, and support from upper management with a specific and visible champion of collaboration. However, some participants were not sure that three-way collaboration was needed. In contrast, a few indicated that they thought more stakeholders needed to be engaged than were present at the tabletop exercise. Among the questioners, one person succinctly said:

> "We need to establish the need for it [collaboration] and then the purpose for the collaboration."

Action Options

The participants discussed ways to address their concerns, primarily focusing on near-term tangible steps they could take. For example, one person asked his colleagues to "think outside the box" for this discussion; this comment reflected the sentiments of other participants in several regions.

Individuals said that they wanted effective plans to implement collaboration, accurate contact lists for each other, opportunities to learn skills for creating collaboration, and greater understanding of the legal authorities on which their responsibilities were based.

---

* Much has been written about the importance of collaboration and stakeholder participation in environmental health. For example, Beierle and Cayford (2002) reviewed many types of processes and examples of stakeholder participation. However, summarizing the literature on this very broad topic was beyond the scope of this project.

Websites were identified as potential tools to share information with each other and with the public. Two groups thought it might be appropriate to link to one or more hospital websites. Participants in another area suggested developing a protected, statewide site where all water-related personnel could be listed with contact information. Additionally, the capability could be built in for persons listed to update their own information as it changed.

There was a lack of consensus in most areas as to whether a formal agreement or MOU would facilitate better collaboration. One area had such an agreement in place; another area was considering having one. Several groups discussed scheduling regular meetings instead of developing a MOU. Participants in some regions said,

> "I'm not sure we need a formal agreement..."
>
> "Written agreements aren't worth the paper they're on... [pause] ... if they don't get used."

Evaluations of the Tabletop Exercises

Most participants indicated during the evaluation periods that they valued the day of exercises, learned about each other during the scenarios, and benefited from the opportunity to network and identify new ways for working together. Among the 37 evaluation forms returned, 34 participants indicated that the activity increased their understanding of various aspects of collaboration, and increased their ability to serve their community's needs. While many found the exercises well designed and thought provoking, some felt that more specific details and additional partners should have been included. A few said that the activities took too much time, but others indicated that the time flew by and was a valuable investment of their time.

During the Framework portion of the day, the dialogues related to many sections (e.g., the inventory of communication tools and potential partners) were quite extensive as information and insights were shared. Discussions that followed other sections (e.g., legal authorities) were quite brief, due to participants' lack of knowledge or recognition of the sections' potential value. Comments about the Framework indicated that the discussion of principles and goals was the most valuable section, while dialogue about contact information lists was viewed as the least valuable. Several people said these lists already existed, although every region indicated that timelier updating of these lists would be helpful. Sections noted as valuable included the inventory of existing risk communication tools, and identification of potential collaboration partners.

KEY POINTS

The major findings from the field exercises and the related literature review are the bases for the following points. These were key inputs to the final design of the tools presented in Chapter 4.

- Organizations are important framers and definers of public health risk issues.
- Water-related health issues are complex and multi-dimensional; therefore, they require comprehensive attention.
- The key to sustaining relationships and effective decision-making is to check repeatedly on assumptions.

- Especially for complex issues, comprehensive and systematic approaches for identifying and assessing contexts, stakeholders, and decision options increase the likelihood of success. These processes also help participants build skills and uncover assumptions and insights, as well as recognize their shared values and hopes.
- Understanding stakeholders' key characteristics is crucial to grasping that stakeholders merit differing levels of attention under different circumstances.
- In the scenarios, most participants had very limited knowledge of the legal bases of their programs, each other's roles and responsibilities, or what gaps in their knowledge could have adverse consequences in their relationships. They realized in some scenarios that they did not fully understand each other's terminology.
- Participants had narrow conceptualizations of "clinician" and "partner" that limited their abilities to develop effective strategies.
- All regions said they needed better contact lists for physicians, but had never thought to check on the effectiveness of the broadcast fax mechanism for reaching clinicians.
- There was little appreciation of the potential benefits of pre-testing or evaluating communication materials and methods.
- Collaborators viewed stakeholders and their claims in various ways. These differences produced both opportunities and challenges for collaborators to address proactively and strategically in their partnership.
- Core values that nearly everyone felt were essential for collaboration included trust, honesty, and respect.
- The groups developed many ideas for tangible steps (e.g., plans, lists, skill building, etc.) that they could take to strengthen their capabilities to address water-health concerns.
- There was only moderate agreement on whether MOUs or other formal instruments would improve collaboration.
- There was no evidence in the tabletop scenarios that physicians had to be in direct relationships with utilities. Most scenarios indicated that clinicians and utilities valued and used S/LPHAs as bridge communicators. Some participants felt that this function would be successful as long as clinicians are confident that S/LPHA personnel represent their interests appropriately, and S/LPHAs have the resources to serve as bridges.
- Narrow approaches to risk communication options prevented jurisdictions from envisioning effective strategies and achieving optimal outcomes.

CHAPTER 4
TOOLS FOR ACTION

INTRODUCTION

The knowledge gained in this project was translated into decision-support aids. We used the information, views, comments, and perspectives of the study participants (Chapters 2 and 3) and the expertise of our collaborators, NACCHO and AOEC, as the bases for creating two toolboxes for fostering effective collaborations focused on water-related health risk communications.

First, the Framework For Action was designed as a toolbox that utility, local public health agency, and clinical professionals could use in building effective partnerships. The Framework contains a series of phases and steps with supporting tools, rationales and materials to guide potential collaborators through a process of defining, designing, developing, and implementing an optimal partnership for their particular risk communication setting. Because approaches to collaboration must be tailored to fit individual sites and conditions (Magee 2003), it is important to use the tool according to available resources and needs. Steps in the Framework have been created for use on a minimal, intermediate, or comprehensive basis. Although a single Framework was drafted for applications across many different circumstances, there will be a range of outcomes that result from its use. No one outcome from the Framework will be appropriate for all locations.

Second, the Cost-Benefit Scorecard was developed to assist in identifying the best option among a variety of possible collaboration approaches. This toolbox was based on cost-benefit literature (below) and input we received during this project. A representative from one of the participating utilities reviewed an early draft of the cost-benefit toolset, advancing our knowledge of its potential value in the field and contributing to the Scorecard's final form.

GUIDING PRINCIPLES

The American public expects safe, reliable drinking water. Addressing water-related health issues effectively involves a large number of partners, mutually beneficial relationships, complex scientific and policy issues, challenging communication decisions, a great deal of organization, and commitment from top managers and governing boards.

Values noted to be essential to successful collaborations in this project and others typically include trust, mutual respect, and honesty (Chapter 3). Other values such as proactive action, rather than reaction, are often noted. Fundamental principles that shape many collaborations include:

- Systems thinking
- Strategic thinking
- Dialogue
- Shared vision
- Shared success as well as responsibility and risks

Many publications have been written on these principles. There is no question that each is important to successful partnerships. These principles are emphasized in the Framework For Action presented in this chapter. To introduce the Framework, however, discussion of collaboration and its essential underpinnings are briefly discussed.

Collaboration

Most utility, LPHA and clinical personnel in this project stated that they wanted to work more closely together, but needed new ideas about how to do so (see Chapter 2). Collaboration between organizations necessitates examinations of organizational policies and processes. Entities desiring successful collaborations need to appreciate important principles that affect collaboration internally and externally. Many disciplines have researched the characteristics of successful collaborations (e.g., Stokols et al. 2003, Drucker 2001, Magee 2003, Anderson and Jap 2005, Davidson and Waddell 2005, NDWAC 2005, Pike et al. 2005). Principles often reported include the following:

- During a crisis is not the best time to begin developing good working relationships.
- Change is difficult and challenging. Contexts and conditions continue to change, so processes and goals must change as well.
- Collaborators should be willing to look at common issues in contexts larger than their own, and thus be willing to think beyond what is familiar to them. Strong, durable relationships with partners and the public must work within broader community and stakeholder contexts, involving communities and managers of key organizations (e.g., both voluntary groups and governmental agencies).
- Leaders must be aware of changes in political, social, and organizational contexts, and related interests among the collaborators. Effective leaders are attentive to issues that may result in liability concerns.
- Lead organizations (i.e., utilities in this project) should identify and reach out proactively to key partners to improve infrastructures and strategies for monitoring issues and ensuring responsiveness. Partnerships are natural outgrowths of effective communications.
- Successful partnerships rely on local planning, design, management, and institutions. Policies, programs, and practices should be tailored to organizational and community conditions.
- Goals and aims for collaborations must be mutually developed and realistic; they also need to be in proportion to organizational needs and resources.
- The quality of collaborative efforts is more important than the quantity of products; i.e., the quantity will be limited if the quality is not optimal.
- Settings that favor collaboration include leaders who clearly dedicate time, make commitments, contribute their expertise, and trust staff to pursue partnerships. Leaders must exhibit behaviors that are consistent with the degree to which they value collaboration.
- Managers need to identify real and potential constraints and recognize the means to address them.
- Managers and governing boards of directors must understand that the effort necessary to build partnerships requires both time and sustained support. Strong, explicit commitments by organizational leaders and governing boards are crucial to collaboration and program success.
- The creation of underlying trust, relationships, and structures requires proactive partners, regular long-term activities, and recognition that building collaboration is a complex process.

Trust

Trust and relationships are interdependent; trust is the cornerstone for developing collaborative relationships (Blomqvist, Hurmelinna, and Seppaen 2005, Fuqua et al. 2004, Magee 2003, Stokols 1995). Many researchers have observed that trust plays pivotal roles in risk communication and risk management as well (Slovic 2000, Jaeger et al. 2001).

Especially for public goods such as drinking water, trust in the public health sector is essential; partners in public sector collaborations intrinsically share risks (Porter and Baker 2005). The literature and the evidence gathered in this project support the idea that trust must be developed and maintained at every level – individual, organizational, and inter-organizational – to advance collaborative efforts.

In the business sector, it has been suggested that the process of contracting increases trust and the development of collaborations (Blomqvist, Hurmelinna, and Seppaen 2005). Though public agencies do not "contract" in the same sense as business, MOUs or cooperative agreements are comparable models that can be instrumental in building strong relationships (NSWHD 2001).

In the regional exercises (Chapter 3), participants were particularly concerned about not being "blindsided" or left without support of their partners when problems occurred. In the March 2004 workshop (Chapter 2), participants stated that problems among partners need to be addressed promptly and honestly to ensure that trust and credibility are not damaged. The concerns obtained in both settings were expressions of desires for trust and relationships in which responsibilities and risks are shared.

Relationships

Participants in the March 2004 workshop and Winter 2004-2005 regional exercises consistently stated that positive relationships, built over time, were essential to successful collaboration. Their views are consistent with professional literature about relationships in partnerships. The role of relationships in collaborations in the business sector are equally valid for the public sector (Anderson and Jap 2005), including those involving water utilities and public health. Review of the vast literature on this topic, however, was beyond the scope of this project.

Relevant here is literature related to public sector relationships. For example, four essential partners for creating, launching, and sustaining public health partnerships have been noted to be governmental, business, academic, and non-governmental organizations (Magee 2003). While individuals and organizations involved in collaborative efforts must purposefully create and develop relationships, they must also attend to power, equity, and other interests to achieve shared goals (Mitchell, Agle, and Wood 1997).

Further, in our research, utility representatives noted that their relationships with S/LPHAs vary, because some health agencies have regulatory authorities over utilities and some do not (Chapter 2). This characteristic of S/LPHAs was the one most often noted as influencing the nature of utility-PHA relationships. When regulatory authority was vested in the S/LPHAs, utilities said that they could not be as open or trusting of S/LPHAs, as they would be if the S/LPHA were not regulatory.

Structure

Structure permits a collaboration to advance from trust and relationships into mutual understanding and action on shared issues of concern. Structural components of collaborations influence partners' abilities to create active, high quality public health programs (Jasuja et al. 2005). Additionally, structure is important from the beginning of the process to create a collaborative forum, and to sustain it and related programs over time (Pluye et al. 2005). Some jurisdictions have found formal instruments, such as Memorandums of Understanding, and work plans, to be valuable means for sustaining relationships as crises and changes occur (Chapters 2 and 3). Other crucial aspects of collaboration have noted as organizational capacity, processes, and oversight (Magee 2003).

As noted by NDWAC, effective collaborations involve comprehensive planning, implementation through budget commitment, and continuous monitoring for improvement. Partnerships should have clearly stated authorities, roles, responsibilities, expectations, and protocols. Evaluations that regularly include employees, partners, and customers in two-way interactions are crucial for ensuring continuous improvement (NDWAC 2005).

Programs must have measurable goals and specific timelines. Collaborations change over time, however, requiring changes in measures of progress (Rhoten 2004). Measures may be simple at first and develop over time as the program develops (NDWAC 2005). Process benchmarks – derived from the structure of collaborations (e.g., number of organizations that attend each meeting, frequency of meetings, etc.) – are important tools for evaluating the group's success in moving toward mutual aims and objectives (Claiborne and Lawson 2005). These measures are important to meet accountability needs, but outcome measures to assess progress toward long-term objectives (e.g., reduced number of cryptosporidiosis cases) are essential as well (Rhoten 2004).

Influential Characteristics

Key factors that help or hinder efforts to develop partnerships, whether for programmatic or research collaborations, were found in the literature and are summarized in Table 4.1 (Magee 2003, Stokols et al. 2003, Rhoten 2004, NDWAC 2005).

Guidance for Water Utilities

Finally, insights about collaboration were also obtained from homeland security literature. One source of practical guidance for utilities is the report of NDWAC's Water Security Working Group (NDWAC 2005). This report contains many points that are as relevant for water-health collaborations as for security issues. Some points have been noted above, but additional ones include the following:

- Ad hoc relationships are no longer sufficient to identify and address emerging problems. Sharing information, planning together, and coordinating responses are important at all organizational levels.
- Establishing formal agreements (e.g., MOUs) may be necessary to ensure regular exchanges of information, clarification of roles and responsibilities, and proactive needs assessment and responses.

Table 4.1
Factors that affect collaborations

Factors	Foster collaboration	Hinder collaboration
Experience	History of joint work among the partners	Lack of experience with successful decision-making processes
Structure	• Cohesive core group involving several staff levels with – Diverse expertise – Clear commitment • Evaluations for continuous improvement	• No clear leader, convener • Changing participants
Interactions	• Series of introductory information-sharing meetings • Informal socializing • Frequent informal face-to-face meetings • Brainstorming sessions • Retreats	• Over-reliance on e-mail as the communication modality • Too few face-to-face interactions to sustain relationships
Common ground	• Unifying vision • Common principles, values • Shared problems • Mutual goals • Jointly developed framework for addressing problems • Clear measures of progress	• Insufficient or unrecognized benefits of collaboration • Lack of tangible progress
Operations	• Physical proximity • Dedicated resources • Flexibility to adapt to change	• Insufficient resources to achieve goals
Time	Sufficient time to • Build trust and personal relationships • Appreciate each other's perspectives and personal styles	Lack of patience with the time needed to • Build working relationships • Resolve problems early on • Ensure success

- Fundamental issues must be addressed in an informed and systematic way, considering existing and anticipated conditions.
- Security (in this project, collaboration) should become part of the organizational culture and daily decision-making processes.
- Prevention and preparation reduce the potential and impacts of adverse events, and thereby contribute to the public's confidence in responsible organizations.

Collaborations based on sound principles and guidance are more likely to produce tangible successes and thereby become increasingly valuable to their partners and stakeholders. Many of the characteristics of successful collaborations are closely related to fundamental elements of strategic decision-making processes.

STRATEGIES

Strategies are systematic processes comprised of plans, methods, and a series of actions designed to balance conflicting interests and issues, and produce specific results. Key components of strategies are:

- Definition of the problem
- Statement of goals
- Assessments of the situation
- Information gathering
- Identification of resources
- Determination of the options
- Selection of the best option
- Implementation
- Evaluation

Strategies tend to be long-term and large-scale; they may serve as frameworks for responses to specific issues that are not known when the plans are first prepared. For example, as part of its chemical pollutant risk management strategy, the USEPA adopted the National Research Council's risk assessment framework (NRC 1983). Initially, it was used to address specific Superfund site issues, but over time the framework was applied – not only by USEPA but worldwide – to address national scale drinking water, air, and other environmental and occupational issues.

But developing effective strategies is often difficult; like collaborations, they require deeper grounding in the scientific theories, methods, and findings about strategic and systems-oriented approaches to risk management. This is especially true for managing and communicating about issues as complex as health concerns related to drinking water contaminants, and even more so for emerging contaminants. The major findings from literature on strategic planning and systems-based approaches were used as important inputs for designing the two toolboxes below, the Framework For Action and the Cost-Benefit Scorecard.

FRAMEWORK FOR ACTION

The basis for the seven-phase Framework For Action requires ensuring that three essential components are in place: trust, relationships, and structure. All three are interrelated; one cannot produce collaboration without the other. However, weakness of any one does not preclude working on the Frameworks' components in a deliberative, stepwise, or iterative manner. In other words, creating an effective collaboration is not an all or nothing proposition; it is a process. Addressing each component actually assists the development of the collaboration. Important results of working on the Framework's components are increased trust, stronger relationships, and implementation of actions made possible through the sharing of resources.

Elements of the Framework

We then defined the criteria that had to be met to create an effective Framework. First, the tools and guidance that comprise the Framework For Action had to result in tangible outcomes based on shared vision and values, and clear goals. Further, objectives had to be specific, measurable, easy to use, and informative. A logical sequence of elements had to be compiled, based on "best practices" from the business literature and lessons learned in the data gathering components of the project. Each tool had to be clear, relevant, and fairly easy to use. At a minimum, the tools had to help each organization identify:

- What their mutual needs are,
- What principles and vision should be the basis for a multi-party collaboration,
- Who is responsible for which risk communication activities,
- What resources are available,
- What related authorities and legal instruments are in place,
- Which organizations are current partners,
- Which organizations may become partners,
- What actions are needed to establish collaboration, and
- How soon desired outcomes could be achieved.

Ideally, the process of compiling the information for each element of the Framework would also help partners realize assumptions and incorrect knowledge they have about each other, provide them with an opportunity to share correct information, develop common ground for expanded relationships, and lay the foundations for effective collaboration.

Further, the Framework needed to be structured so it could be used according to the level of available resources and commitment; it has been designed for use on a minimal, intermediate, or comprehensive level. For example, where a utility has limited personnel, a basic approach may be appropriate. An intermediate use of the framework may fit organizations that have some elements of collaboration in place with its partners, while comprehensive use of the tool may work best for entities that have several staff members who can focus on building and sustaining collaborations with S/LPHA personnel and clinicians.

Additionally, the Framework needed to be designed so that it could be used incrementally; e.g., regions can step through the process one element at a time, rather than complete the entire set of tools at one time. In fact, using the Framework in a stepwise manner provides organizations with opportunities linked to a structured process for gradually building knowledge and experience with each other over a period of time.

Interaction is the most basic of requirements for communication and collaborations (see Table 4.1). Regular communication should occur as frequently as will meet the mutually agreed upon goals and available resources. Explicitly committing to and conducting regular communications fosters collaboration. These interactions may utilize a combination of routinely scheduled conference calls, e-mail updates, protected websites (e.g., id-access chatrooms), or other forms of communication. However, there is no substitute for "face time" (Stokols et al. 2003, Rhoten 2004). Throughout the collaboration process, a structured series of orientation and information seminars, breakfasts, brainstorming sessions, or a multitude of other interactive formats can be used to share information, establish relationships and trust, and lead to creative discussions and insights.

Table 4.2 presents the overall structure of the Framework For Action and identifies the tools that link to specific steps.

Table 4.2
The structure and tools for the Framework For Action

Phase	Step	Tools
1. Prepare the process	1.1. Determine the need	F1.1
	1.2. Identify partners	F1.2
	1.3. Plan the process	
2. Start the dialogue	2.1. Convene the partners	
	2.2. Define the problem	(Review F1.1)
	2.3. Decide logistics	
3. Deepen the dialogue	3.1. Develop a vision	F3.1
	3.2. Agree on underlying values	F3.2
	3.3. Define the types of information needed	F3.3
4. Learn from current conditions	4.1. Conduct inventories, gather information	F4.1.1 F4.1.2
	4.2. Assess resources, mandates, authorities, roles, performance, trends, etc.	F4.2.1 F4.2.2 F4.2.3
	4.3. Develop insights about current conditions, contexts	
	4.4. Share information and insights	
	4.5. Identify gaps, overlaps, and opportunities	
5. Create new ideas	5.1. Identify strategic issues	F5.1
	5.2. Recognize why these issues are strategic	F5.2
	5.3. Sequence the issues	F5.3
	5.4. Define the goals	F5.4
	5.5. Develop options	F5.5
	5.6. Select the best option	
6. Plan to take action	6.1. Plan the process	
	6.2. Define the action plan	
	6.3. Define measures of success	
	6.4. Obtain resources	
7. Take action	7.1. Implement the plan	
	7.2. Evaluate	
	7.3. Feed findings back into the action steps	

Phase 1: Prepare for the Process

The first step in the process is internal to the initiating organization. In this case, we assume that the water utility is the lead organization that has an interest in exploring the possibility of new or stronger collaboration for addressing water-health risk communication. The Framework has been prepared on this premise, but may be readily adapted for use by a local public health agency or other interested partner.

Step 1.1: Determine the Need

First, the lead organization needs to recognize the value of initiating a process to develop or strengthen collaboration. This may occur because of problems during a recent crisis event or emergency drill, a new legal mandate that has been placed on the utility, a change in organizational leadership, new societal contexts, or some other change. Consider whether the related issues are strategic to the organization:

- Are they issues that must be addressed to achieve the organization's vision?*
- Are there new opportunities that should be utilized to position the organization and/or community more proactively for addressing future or emerging issues?

Whatever the initiating cause, utility staff, the general manager and/or the governing board must recognize that action is necessary to address the issues.

One of the most important parts of developing a strategy is defining the problem, which may begin with discovering that collaborators and/or stakeholders view the problem in more than one way. When that is the case (as it most often is), it is crucial to develop a mutually "owned" and explicitly stated problem; e.g., a reframing of the perceived set of problems so that all partners agree on what the problem really is. Finally, definitions of challenges related to the shared problem should be informed by scientifically derived data (Magee 2003).

A core group within the utility may draft a statement of the problem(s) they see, and a proposal to address the challenges. Answering some major questions can serve as the stimulus for this draft statement (**Tool F1.1,** beginning on page 79, all tools are presented sequentially). When concurrence about the need is in place across organizational levels, a final statement of need can be prepared. Upper management should not only be aware of the statement, but also visibly support staff efforts to proceed with determining the interest of potential partners.

Any common ground shown in Table 4.1 (e.g., shared vision, values, needs, etc.) can serve as a starting point for developing a statement of purpose for the collaboration (step 1.1). Without a firm agreement on the need for collaboration, however, there is little likelihood of sustaining a group focused on drinking water-related health risk communications or other issues.

---

* Note that vision is addressed later in this process (Phase 3, Step 1) among the collaborators and is distinct from an organization's mission.

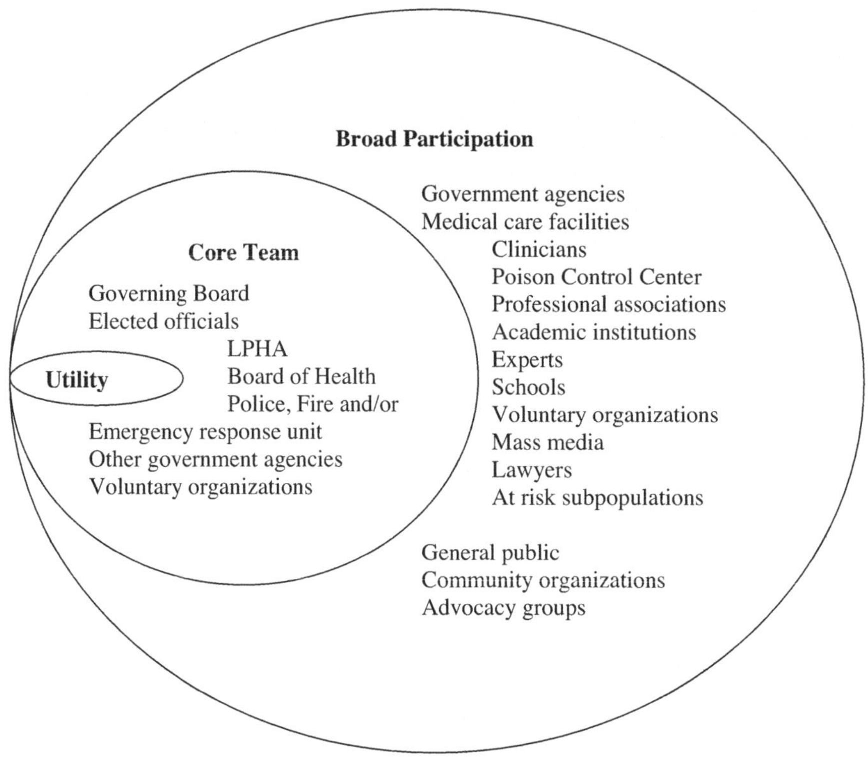

Figure 4.1 An example of core and broad sets of partners

Step 1.2: Identify Potential Partners

Key steps in the process are identifying potential partners and engaging them in the process. This step focuses on the identification of potential partners. In the regional workshops (Chapter 3), all groups had narrow views of "clinicians" and "partners." These limited concepts constrain the participants' abilities to identify potential partners. As a result, we designed this step to help users expand their thinking about what organizations should be evaluated as potential collaborators.

Partners need to be relevant and interested in the problem identified, as well as available and willing to collaborate. As potential partners are identified, their issues, needs, priorities, expectations, constraints, and concerns also need to be recognized; these may need to be addressed upon first contacting the candidates.

Partners within the utility (and S/LPHA) should be identified, as well as organizations outside of the utility. For example, if the Director of Water Quality conducts this phase of the Framework, internal partners to consider may include the public information officer, the director of community relations, the director of customer services, the general manager, and the board of directors.

In starting the process, one of the most important decisions is how broadly "potential partners" should be defined. Does the identified need call for a broad representation of interests, expertise, perspectives, etc.? Some jurisdictions have started widely (see example in Appendix B), while others have begun narrowly; e.g., with LPHA and mayoral staff representatives. The decision should fit local conditions and organizational capacity to initiate contacts. In some cases, the utility may decide that the most practical approach is to start narrowly and allow the initial dialogues with "core partners" to point to additional stakeholders to include in the process. (See Figure 4.1) Practical issues related to group size also need to be considered.

Tool F1.2 is based on a widely applied mechanism for recognizing key features of stakeholders and using these characteristics to determine the importance of potential partners (see Chapter 3). This tool is an important starting point for developing collaboration. The user of the tool should remember that, although the perspective of the lead organization drives its application, potential partners might judge their own characteristics differently than the lead organization. These differences in perspective should be recognized at this point and addressed later in the process (e.g., Phase 2, Step 2).

Step 1.3: Plan the Process

With a statement of need and a list of potential partners completed, the next step is to determine whether the organization's resources and commitment are sufficient to proceed in contacting the initial set of partners. If not, one strategy would be to engage partners who are willing to contribute resources to implement the process; e.g., have a facility large enough to convene the group, have a regularly scheduled meeting involving many of the potential partners where the utility's request for collaboration could be introduced on the agenda, etc. In some cases, the General Manager or members of the governing board may play key roles in reaching out and engaging the interest and resources of potential partners.

Several things are crucial before advancing beyond this step:

- The support of upper level management
- Sufficient resources
- More support than opposition within the lead organization
- Core staff are willing and enabled to implement the process
- General agreement on the need, scope and preferred process

Planners should also develop an estimated budget and timeline for the process. This initial planning and engagement phase may take a few months, but the entire process to reach the action phase may take a year or two. However, the time needed is highly related to local needs and conditions, e.g., urgency, relevance, resources, etc. Potential partners will want and need to know what kind of time and level of commitment they should expect if they agree to be involved.

If the organization is not ready to proceed into a full process, perhaps some elements can be implemented or the scope of the problem limited to a more feasible scale. A clear statement of how the organization will proceed should be documented in a proposed work plan that has broad support within the organization.

Phase 2: Start the Dialogue

The potential partners are then invited to engage in the process of exploring the possibility of collaboration to improve water-health risk communications. Several conversations may be needed with some partners to hear and address their concerns, strengthen their interest, and obtain their agreement to attend at least the first meeting.

Remember that dialogue is a fundamental principle of collaboration. Creating means of listening, dialoguing, and addressing the needs of all partners are pivotal to sustaining the initiative.

Step 2.1: Convene the Partners

Establish a mutually convenient time and place for the first meeting of the core group. The agenda for this meeting should include discussions of: the risk communication problem(s) identified, the scope of the problem(s), statement of the problem that everyone in the group can agree on, the need for collaboration, the scope of the process needed to address the problem(s), and the next steps.

If participants do not know each other, it is important to take a few minutes for each person to introduce him/herself, areas of responsibility, and interest in the issue. If participants do know each other, a discussion of how each individual has experienced the problem could be used to identify points of common concern.

MAPP (Mobilizing for Action through Planning and Partnerships), a strategic planning and performance improvement tool for local public health systems developed by NACCHO (http://mapp.naccho.org/strategic_issues/index.asp), provides additional tips for strengthening new partnerships.

Step 2.2: Define the Problem

Several methods can be used to mutually define the risk communication problem to be addressed. For example, participants could decide that the problem is poor communications between agencies during emergency operations or drills, and then address that specific issue. Or the group might decide to tackle communication collaboration broadly; i.e., develop formal foundations and mechanisms to address water-health risk communications on an ongoing basis. In either case, stronger, more organized relationships between partners will be recognized as a need, and responsibilities and resources can be more readily shared.

Regardless of how narrowly or broadly the problem is defined, the process of defining the problem must be inclusive. All participants need to have the opportunity to be heard and to have input in the final problem statement. Depending on the group's familiarity with the triggering factors (of what) and each other, each participant could use **Tool F1.1** to prepare a one-page statement of need in advance of the meeting. These drafts could then be used in the meeting as a starting point for identifying areas of common and differing concerns.

The product of this step should be a succinct, mutually accepted statement of the problem. The scope of the problem should be clear so that the following steps can be implemented appropriately.

Step 2.3: Decide the Logistics

Once the problem is defined, the group should decide how to continue the process. Issues to address include how often and where meetings will be held, what means of communication will be used between meetings, whether a central communication point (e.g., chat room, protected website, or other method) will be used to advance the groups' thinking and work.

Additionally, the group should determine whether other potential partners should be invited to participate in the collaboration. The scope of the problem may have been defined in a way that would affect or require the involvement of groups not initially considered (e.g., schools, nurses, dieticians, hospital communication personnel, etc.). The group could use **Tool F1.2** for identifying and prioritizing potential partners. Each participant could use the tool between meetings, and the group could use their assessments to decide which potential partners should be invited and what they would add to the Framework process.

Serious consideration of how open the process should be is appropriate at this step as well. If the outcomes of the collaboration will have visible impacts on the public or at-risk subpopulations (e.g., rather than a collaboration focused on improving interagency operations), representatives of those groups should be engaged in the ongoing dialogue. Obtaining their input early on can help keep the decision-making process focused on its ultimate public goals.

Once the what, when, and who of the process have been mutually agreed, the partners need to determine what resources (e.g., financial, personnel, staff skills, technology, meeting space, etc.) will be needed for the group to proceed. Locating and obtaining the necessary resources will probably require work outside of group meetings. Each partner should be encouraged to provide some form of support to the process, but should not be excluded from the group if tangible resources cannot be identified.

Phase 3: Deepen the Dialogue

The purpose of this phase is to ensure that all partners share a common vision of where the process will lead and what its results will be. As was found in the regional workshops (Chapter 3), individuals often assume they know what others think and believe about an issue or problem. However, an effectively facilitated discussion of what everyone wants to achieve inevitably provides group members with new insights about others' perspectives and concerns, and about where their own views fit in the group as a whole. Mutually creating the vision not only helps to build relationships and knowledge within the group, but also sets the overarching goal for the process. This Phase ensures that all participants have the same opportunity to be heard, to shape the vision, and to contribute to the fundamental values that will serve to underpin the entire process.

One of the most important aspects of this Phase is the neutrality of the convener and the setting in which these dialogues occur. Careful consideration should be given to either hiring an external expert facilitator, or enlisting the support of a neutral and skilled leader in the community or adjacent community. These discussions should not be dominated by the utility or organization that started the Framework process. Even holding the meetings in a place that reinforces the importance of the convener can create stress for other participants (e.g., large organizational crest on the podium, strong presence of the organization throughout the location or meeting's events, non-participating staff frequently coming into the room, etc.). Ensuring that no one organization dominates this Phase is essential.

Other key aspects of this Phase are engaging the participants, expanding their knowledge and views of each other, energizing them to set higher standards, deepening their commitment to the process, and motivating them to reach for better performance both individually and collectively. The strength of the bonds they build during this Phase will yield comparable benefits throughout the Framework and beyond.

Step 3.1: Develop a Vision

The outcome of this step is a brief statement of the group's ultimate goal for the process. It captures how they want to see communication occurring at a defined point in the future. The collaborators must determine what the focus will be and what their ideal solution to the problem (defined in Phase 2) looks like. Their vision of the solution must be both inspirational and motivational to sustain the process and inform others about what they are striving for. Their thinking needs to be freed from the current situation so they can be creative and brainstorm a vision of how things will work and be when the problem has been solved. A neutral, skilled facilitator can assist partners in shaping their vision by providing open-ended questions (see examples in **Tool F3.1**).

The scale of the vision will relate to the scope of the problem. A narrow scope may require a more limited and shorter-term vision, while a broad scope may require envisioning interagency communication capabilities and performance five to ten years in the future.

If partners have conducted visioning exercises before, they may bring their experience to this step to contribute to the process. However, while effective facilitators build on these past experiences, they do not allow them to become roadmaps for this step. The visioning process needs to be shaped by the participants, driven by their priorities and issues as well as their desires to achieve communication improvements. Participants may need to be reminded that no two visioning exercises are alike. Past conflicts and difficulties do not need to be addressed or resolved in this phase, but they should be acknowledged and respected as part of the participants' collective history. Most importantly, negative experiences should not be allowed to constrain the brainstorming exercise. There are many techniques that can be used in this step, but describing these is beyond the scope of this project.*

Step 3.2: Agree on Underlying Values

Values are the principles and beliefs that define and guide the group process. They set out how the participants want to interact, and what they expect of each other and the process. Some groups and facilitators prefer to identify underlying values before defining their vision, while others find it too difficult to state their values until they have a mutually agreed vision. Either way, the values that the participants share provide the essential foundation for the entire Framework. When the group's energy lags during the process, or their focus strays from the stated problem, values can be used to bring the group back to its central purpose and ideals.

Collaboration depends on participants believing that they are a team, and behaving as if they are a united set of people. In the words of one regional workshop participant, "We are in this together." Others in this project expressed concerns that they would be "blindsided" or hurt by each other if they did not address their differences and start working together on solving risk communication problems. Partnership requires participants to recognize that the interests of others must become their interests, that they share responsibility for all aspects of the collaboration, that they must bring assets to the endeavor to sustain their place in the process, and that they must help each other when problems occur.

Some of the basic values that support collaborative efforts, relate to the behaviors that participants expect of each other (e.g., trust, respect, honesty, timeliness, commitment, etc.) and

---

* See http://mapp.naccho.org/referencesandresources.asp and http://mapp.naccho.org/visioning/ for suggestions. A skilled facilitator will know about many suitable options.

principles of good practice in the public service sector (e.g., prevention, leveraging resources, building community relationships, etc.). (See **Tool F3.2** for questions that can assist in recognizing values.)

Once the group has identified its common values, the values need to be written in a brief, simple statement that is acceptable to all participants. This statement can be included in materials, meetings, and other ways to keep them visible to the group and stakeholders, and thereby help the group stay focused on what they agreed is important to the process. Further, the group may wish to share their vision and values with potential partners to enlist their support for the process and obtain their enthusiasm for the intended outcome.

Step 3.3: Define the Types of Information Needed

In a separate meeting or process, the group should determine what *kinds* of information need to be emphasized during the next phase, focused on assessment. (**Tool F3.3** lists some major categories usually required for assessment.) One step in this process is to assess the types, availability and characteristics of information materials internal and external to the organization.

Participants must be fully informed about, understand, and then learn from current conditions before they can identify what the underlying causes of their communication problems are. In most cases, the multitude of forces and factors that affect these efforts are too complex for any one individual or organization to have complete knowledge of all the contexts in which communications occur or how problems arise. A process of obtaining the same types of information for each participating organization and bringing the information to inter-organizational discussions (in Phase 4) is essential for developing new insights about current constraints and assets that can be leveraged.

Assessment is inherently a time-intensive and iterative process; it is rarely completed in one round. To begin, the participants need to have a common understanding about the broad types of information needed. They will need to determine how they will gain access to that information, whether they will have to reshape it for Framework purposes, or resolve other logistical issues in preparation for Phase 4.

Phase 4: Learn From Current Conditions

Communication problems require comprehensive, system-wide thinking to identify the factors that are responsible for important changes in the societal, cultural, legal, organizational, and other contexts that coincide with the issue. Communications occur on many levels, from interpersonal to cultural domains. When communication activities are based on a narrow vision of the communication environment (e.g., often the individual scale is the only one considered), optimal outcomes are not likely to be achieved. It is crucial to gather information on multiple levels (individual, community, and societal scales) and from different perspectives (e.g., legal, cultural, organizational) in advance of crises, so that a foundation of broad and widely accepted knowledge about the communication environment can be utilized as problems arise.

This phase involves assessment of current contexts and components of communication and risk communication activities; it provides the essential information needed to identify the strategic issues that must be addressed to achieve the collaborators' shared vision. This phase requires the review of existing drinking water-health risk communication materials, consideration of existing partners, and relevant mandates, authorities, policies, programs, materials, and services. In other words, both the "*what*" and "*why*" of current conditions need to be examined. The knowledge gained from this important phase is often used to return to and refine the problem statement, and then reposition the partners for effective action. This kind of feedback loop may occur at any point in the strategic process, but is not uncommon at this point or after resources and limitations have been recognized.

Water-health risk communication services and factors may be in different organizations to meet the needs of different stakeholders or to accomplish differing purposes. A tangible benefit of this phase is the compilation of the current elements that relate to water-related health risks. From this compilation, overlaps and gaps in services and materials can be more readily identified, and insights developed about the jurisdiction's needs and opportunities for improvements.

It is very unlikely that any one participant or organization will have all of the information necessary to compile the full context and trends of an area's risk communication activities. This is one of the reasons why a collaborative process is so essential to addressing risk communications. Utility, public health, clinical, and other personnel all have different perspectives, knowledge, and information; each is important to understanding the components and interrelationships of the water-health risk communication environment. The value of implementing the Framework expands with the breadth of input brought especially to this phase of the process.

Any problems that collaborators may face in either information collection or sharing should be discussed early in this phase. Barriers to sharing information will constrain the Framework process. Without adequate information as the basis for later steps, the Framework cannot produce optimal outcomes. Fundamental choices in this step will affect the value and impact of the remaining Framework processes.

Jurisdictions have differing types and levels of risk communication services, and different levels of resources available for conducting a survey of current conditions. It is important for collaborators to develop mechanisms that are sufficient for fully informing the group's decision-making process without overwhelming any organization's capacity to gather the information requested. In some cases, recent information and data gathering may be relevant and useful without duplicating the effort in this step. These sources of information should be considered as the team determines how the inventory step should be conducted. While comprehensive information is the ideal, partners should be asked to gather no more information than their organizations can readily provide.

There is no one right way to conduct this phase. However, using a systematic and standardized approach, acceptable to all partners, is important to ensure that the information brought back by each participant can be readily synthesized and understood across organizations, and thereby more rapidly produce new insights among the collaborators. Some jurisdictions may decide that these tools can be completed at the same time by different staff; others may wish to focus the data collection process to a limited number of people and work through these steps in sequence. Nonetheless, the decision about sequencing or not is less important than choices made about the quality of the information to be gathered.

Step 4.1: Conduct Inventories

To understand what risk communication services exist, partners need to gather information within their organizations and then share their findings among the collaborators. Data collectors need to take care in accessing information throughout their organization, and avoid assuming without question that they already know where programs and materials exist. This problem was noted in the regional exercises conducted in this project (Chapter 3); frequently people working in the same agency were not aware of each other's risk communication materials and services. Units not represented in these workshops may have had relevant programs that were not identified in the regional discussions.

Additionally, many participants could not readily identify their current communication partners or what services they provided (Chapter 3). Upon reflection, participants began to realize that schools provided space for public programs, the Mayor's staff conducted public outreach and media services, etc. A cross-organizational inventory of partners may help collaborators recognize partners who can serve in roles that no one organization would have realized. Recognition of such partners is essential to building a comprehensive picture of the contexts in which risk communications occur, successes are achieved, and problems arise. Additionally, collaborators may want to evaluate their partners' prominence by using the Mitchel, Agle, and Wood (1997) paradigm presented in Chapter 3.

The inventory tools we present (**Tools F4.1.1 and F4.1.2**) are designed to help standardize data collection. The tools can be implemented to the extent that teams are able; e.g., the left-hand columns are typically the least demanding level of information, while the right-hand information may be the most challenging to obtain. Before teams begin conducting their inventories, collaborators should agree on the minimum level of information needed for effective implementation of the Framework. Some teams may wish to complete the first two or three columns before deciding whether additional information should be sought. Further, collaborators may recognize questions that need to be added to the tools so that the level of information needed in their context is sufficient for decision-making purposes.

This step can be completed in increasing levels of detail as the group's decision-making requirements deepen. If the data collection process is truncated without obtaining meaningful insights, however, the value and outcomes of the Framework will be limited. Therefore, the user is advised to complete the columns in the Tools as far to the right-hand side as possible.

Step 4.2: Assess Bases, Resources, and Trends

Another finding from the regional workshops was that many participants were not familiar with the legal or other bases for drinking water-related health risk communications (e.g., for boil water notices – calling them on and off; see Chapter 3). However, a more informed basis to understand each other's roles and responsibilities was also found to be important especially to utility and LPHA personnel (Chapters 2 and 3). As a result, we designed a step and tool (**Tool F4.2.1**) to assist teams in examining the foundations and contexts for the services they provide. Examples of questions to guide this assessment are:

- What are the reasons these materials and services came into being?
- Does a law, regulation, or an executive directive mandate these services?
- Were they developed to meet growing stakeholders' needs and demands?
- Were they created proactively in anticipation of emerging needs?

Another important aspect is the determination of resources each partner has available, or can make available, to address their shared risk communication problem (**Tool F4.2.2**). The type and level of resources required to sustain these services also need to be understood, not in exact or audited terms but in a more qualitative sense.

The resources do not have to be the same in nature or extent, but each partner should be willing to articulate the type or estimated set of resources that are relevant to the problem and potential solution. It is important as well for partners to identify what they cannot contribute to the collaboration; e.g., limitations of staff time, funds to print written materials, authority to handle certain aspects of the problem, etc. It is just as important for partners to clearly state what they can do as well as what they cannot do or provide. Such frank information sharing helps build the climate of trust and openness so essential to successful partnerships.

Questions to consider, for example, are:

- Are resources sufficient to continue these services?
- Are some services or materials no longer needed and therefore ripe for discontinuation?
- Can resources used for duplicative approaches be redirected into complementary or missing services, or are the resources "blocked" for these services only?

Further, what has been the past performance of risk communication activities?

- What have been the impacts of the services provided?
- Have the services performed as expected?
- Are the services valued by the end-users?
- Are the services making enough of a difference to continue as they are, or should they be modified?

Routinely conducted customer surveys, telephone logs, or other ongoing data collection may offer insights about trends in customer satisfaction with services and the intended impacts of services (**Tool F4.2.3**). Also,

- Are there demographic, economic, political, societal, risk perception, or other changes occurring?
  - Could these trends influence the nature and value of risk communication activities as they are now conducted?

Comprehensively understanding the contexts in which the services are provided is essential to conducting a fully informed analysis of current conditions.

Step 4.3: Develop Insights

Using the inventory of current services and additional information about mandates, performance, resource commitments, and key contextual factors, participants should be able to begin developing new insights about their own organization's and other's risk communication services, conditions, and issues. Examples of questions that can lead to organization-specific insights include:

- Are there overlapping services or redundant materials that can be shared and leveraged?
- Are there important gaps in services?
  - Are key or at-risk population groups not being served?

  - Is any organization already charged with providing the missing services? If so, what are the current reasons for the gaps?
- Are resources sufficient to keep current services updated with scientific and regulatory trends and meeting stakeholder needs?
- Which services are producing the intended outcomes?
  - Which are not? Should these services be revised or discontinued?
- Are there growing concerns among the general population or specific subpopulations that need to be addressed?
  - Are these concerns specific or narrowly focused?
  - Are these concerns linked to specific subpopulations or broadly held?
- Are at-risk populations adequately informed and able to act appropriately to prevent water-related illnesses?
  - Who are the at-risk subpopulations? Can they be readily identified? If not, why?
- Are clinicians able to address patients' drinking water related questions?
  - Do clinicians know where to get the best information to address such questions?
- Are new regulations expected that may require a new risk communication campaign?
- Is there new political or managerial leadership?
  - Does the leadership have new expectations of how the public's concerns should be addressed?
  - Does the leadership have more or less interest in water-related health issues?

These questions are not intended to be comprehensive, but point to some areas where new insights can be gained by compiling information across partnering organizations. Often one organization's daily experiences may seem too trivial to share, until the same issues are expressed by one of more partners. The emerging theme may point to a larger issue than any one organization had realized, and call for a larger scale response than any organization could have provided. Such a finding could point to a strategic issue to be addressed in Phase 5.

Step 4.4: Share Information and Insights

At a designated point, all partners should share their completed inventories with their collaborators. The participants should have an open dialogue about their ability to complete the tools to the level set by the group (in Step 1 of this phase), and about the problems they encountered. The participants need to state the assumptions made when gathering their organization's data for this step and how the assumptions affected the quality of their inventories. All partners need to know whether the available data are comparable across organizations; understanding assumptions are crucial to make this inter-organizational assessment of data quality.

The opportunity for doubts and the potential for distrust must be minimized promptly. Open dialogue about concerns is crucial before the evidence is evaluated; it helps build empathy and understanding about each other's unique perspectives and contexts. Some of the most important outcomes of implementing the Framework are building trusting relationships among the partners, deepening collaborators' sense of shared responsibility and ownership for risk communications, and developing capacity to address problems more creatively and effectively. These intangible outcomes merit as much careful attention as the data gathered. If tensions arise, partners can return to the shared values they established earlier in the Framework.

To obtain insights about the partners' risk communication environment, many of the questions in Step 3 can be asked in this step, but they are answered using data from all participants. Looking across organizations, consider additional questions such as:

- Are there new or emerging public concerns that more than one organization is hearing and/or handling?
- Will new mandates for one organization place demands on others to provide related services?
  - Are these partners aware that their services will be required?
  - Do they have the capacity to respond?

Finally, the collaborators may want to review the collective prominence of their partners under both routine and emergency conditions (using the Mitchell, Agle, and Wood (1997) typology presented in Chapter 3). The results of this analysis should help clarify which organizations not included in the Framework process are most likely to be key partners and why. At this point, such partners could be invited to join fully in the process and collaboration.

Step 4.5: Identify Gaps, Overlaps, and Opportunities

Building on their new insights, collaborators can more effectively identify the existing gaps and overlaps in services, as well as the opportunities for improvements. The growth in the group's insights will continue past this step, but a focused discussion of limitations and the potential for change is important at this point. The knowledge gained and creative thinking based on that knowledge will provide key inputs to the next phase, which involves identifying strategic issues, defining goals, and selecting the best option for achieving the group's vision.

The discussion is this step may generate richer results if a neutral and skilled facilitator leads it. When the participants do not feel that one of the group's members is "in charge" of making sure new insights are found or "marking every word," freer expression and dialogue is likely to occur. Some partnerships will not feel that they need outside support to extract the range of insights they should in this step, but others will find this option appealing.

Phase 5: Create New Ideas

The shared knowledge and insights gained in the Assessment phase (Phase 4) provide the basis for this series of steps, focused on identifying innovative ways to address the key problems. At this point in the Framework, participants have a clear statement of their shared vision and a clearer understanding of what causes their communication problems and challenges. A key component of this phase is recognizing which contributing factors merit earlier and/or the most intensive attention. Among the many factors there are some which, when addressed, will in turn resolve other problems or make solving them easier. Identification of strategic issues is essential to this phase. Further, to achieve their vision, the collaborators must develop a range of options for addressing the strategic issues and then determine which option is optimal.

This phase may require the support of a skilled facilitator to ensure that participants' thinking does not revert to traditional approaches, especially those that do not fit current conditions. Participants need opportunities to "think outside of the box" and feel free to suggest whatever comes to mind; unbounded thinking will open up innovative ideas and approaches. Even if an option is discarded because it seems unrealistic, it should remain available to the group; there may be valuable elements in the option that could be used in another more appropriate alternative. Open dialogue and creative interaction is fundamental for crafting new approaches; such dialogue depends on trust and respectful working relationships. To gain the full value from this phase, ensuring that the group's values are reflected in this Phase is essential.

Step 5.1: Identify Strategic Issues

What are "strategic issues" and why is it important to identify them? On NACCHO's MAPP website,* "strategic issues" are defined as

> "... those fundamental policy choices or critical challenges that must be addressed for a community [or collaboration] to achieve its vision."

The site further indicates that strategic issues are more than critical in nature; they also offer pivotal opportunities to look forward, take advantage of current and emerging conditions, and proactively position the collaboration toward its vision. Strategic issues are not necessarily urgent; in fact, the urgency of non-strategic issues draws attention away from strategic issues. Ideally, the collaboration will develop ways to remain resilient so that urgent issues do not drain needed resources away from strategic ones; otherwise, the effort to reach the group's shared vision will be come extended over a longer time period or cease entirely. Focused and sustained effort directed to strategic issues will position the collaboration for future challenges.

Strategic issues can be identified from the information shared up to this point in the Framework process or new inputs from other sources. While a great deal of data will have been shared and many ideas and insights shared, the participants need to sort through what they know and recognize which issues qualify as "strategic." First, the group needs to review its vision and values and then consider the themes that emerged from its assessments. As the MAPP site states, a key question at this point is

> "What factors identified in the assessments *must be addressed* in order to achieve the vision?" [emphasis added]

Often, participants' responses will begin to converge; i.e., their solutions will be contributing a fundamental issue, rather than several issues. This convergence may not be apparent to the discussants at first; a facilitator may need to guide their discovery of this so that the group will become aware of which of their issues are, in fact, strategic.

Another way to identify strategic issues is to consider what the consequences would be of not addressing them. Sometimes it is easier for groups to discuss issues from the negative perspective than positive. Either way should lead to convergence of ideas around the issues that are truly strategic to the vision. (See **Tool F5.1** for example questions for this step.)

---

* See http://mapp.naccho.org/strategic_issues/index.asp for further information.

Step 5.2: Recognize Why These Issues Are Strategic

In the discussion to identify the issues that are strategic, the nature of the issues must be considered. Understanding ***why*** issues are strategic is essential before moving forward in the Framework; this understanding will contribute to the development of alternatives to address the issues.

A key characteristic of strategic issues is whether the issue is directly related to a participating organization's mandates, authorities, services, resources, stakeholders, and/or operations. If the characteristic is directly related, then the issue is likely to be fundamental to the organization and the collaboration's ability to achieve its vision. However, there may be some organizational issues having a key characteristic that are not fundamental to the group's vision. For example, problems linked to the quality of vital records (e.g., birth and death certificates) may relate to a crucial mandate of the S/LPHA, but may not be relevant to the collaboration's vision for water-health risk communication.

Thought-provoking questions can facilitate the recognition of features of issues that make issues strategic. Considering issue-specific features (e.g., urgency, criticality, relationship to organizational mandates, etc.) sequentially may help participants quickly assign importance to many issues. Examining the features that contribute to why an issue is crucial will build understanding of why issues ultimately become identified as "strategic" and thereby open up lines of thought about innovative alternatives for addressing them (See **Tool F5.2**).

As a result of examining *why* issues are strategic, some issues previously thought to be strategic may be eliminated while others previously unrecognized may be added to the group's list.

Step 5.3: Sequence the Issues

One objective of this step is to reduce the above list of strategic issues to about ten specific statements that characterize the fundamental concerns. The listed issues may need to be consolidated into statements that do not overlap or omit significant problems. Short statements of each major strategic issue should be prepared before attempting to order the issues.

Sequentially ordering the list of final issue statements requires consideration of whether the resolution of one issue can affect the resolution of others. Ordering issues sequentially should take into consideration whether there are unique opportunities in the near term to build on, whether the issue is timely or has immediate dimensions, and whether later events offer possibilities for addressing problems. However, some problems may not have a time element; they may offer opportunities for the group to achieve some early and tangible successes that would draw in other partners or add momentum to their initiatives. (Examples of questions to consider are listed in **Tool F5.3**.)

Only in the full societal, political, economic, etc. contexts can issue ordering be considered. As a result, the final list of ***when*** issues need to be addressed will be unique for each partnership.

Step 5.4: Define the Goals

Bridging what *currently exists* to what the group *wants* for the future, requires defined goals; i.e., broad, long-term statements of what the group wants to achieve by addressing the strategic issues. When the goals have been reached, and thus all of the issues addressed, the shared vision will have been realized. Defining goals is a fundamental step in strategic planning and in building collaborations' capacities for improving their risk communication initiatives.

Because of the vast literature on preparing goals and the wide use of goal-setting activities in organizations, most of the participants are likely to be familiar with the tasks in this step. A few key points will be made to enhance what participants already bring to this portion of the Framework.

A few – typically about four – comprehensive, but not overlapping, goal statements should clearly state ***what outcomes*** are necessary for the group's vision to be accomplished. Participants should explicitly link their goals to their vision and strategic issues. By this point in the Framework, participants may have realized that their vision has several components that can be addressed in parallel or sequence. Goals should help clarify what must be done to deal with each part of the vision and with each strategic issue. Goals then could be conceptualized in response to questions (see **Tool F5.4**) such as:

- What are the components of our vision?
- What outcomes must we have to realize any one part of our vision?
- Which outcomes are necessary to fully accomplish our vision?
- What results will we have to have in order to address our strategic issues?
- How do the outcomes we want relate to each other?

The goal statements serve as the foundation for deciding what will be done to achieve the shared vision. Clarity and distinct outcomes are crucial to guiding the decision-making processes that follow.

Step 5.5: Develop the Options

To accomplish the goals, the collaborators must determine ***how*** they will achieve them; this should be done by acknowledging the status quo and exploring a range of strategic alternatives (**Tool F5.5**). Development of options should rely on the type of brainstorming that was done in Phase 2, visioning. To develop new approaches, more "thinking outside of the box" and collective learning that produces creative strategies is essential.

The status quo is the scenario against which the other options will be compared (see Step 5.6 below). Next, a set of potential options must be generated; ranging from those that require minimal change to those that entail extensive, new policies, programs, and partnerships. One option should describe the ideal option, regardless of whether it requires extensive change or not.

The group should have identified no fewer than three clearly different options to consider; these would be the status quo, the ideal (requires the greatest change), and an option in-between these two extremes. A more robust set of options would include at least five distinctly different options, but working with more than five options in Step 6 below is likely to be too time-consuming and resource-intensive. What is important in this step, however, is a comprehensive consideration of a wide range of possible approaches to achieving the vision. This is a key time to think creatively and allow the group to consider all ideas (see **Tool F5.5**).

Once the list of options has been developed, for each option consider questions such as:

- What is the *purpose* of this option?
- What is the *structure* of this option? (Note roles of each organization, linkages among the organizations, and the sources for all necessary resources.)
- What *processes* will be used? (Note the processes for decision-making, leadership, and communication.)
- What is the *time frame* to accomplish the option?
- What *resources* will be needed for this option?

For example, in a baseline scenario (e.g., an informal collaboration for water-health risk communication), roles for each participant may not be explicitly stated, no new resources would be available for the collaboration, joint decision-making may be limited and/or incident-specific, and only as-needed risk communication activities would occur. In a more structured option, formal work assignments are clearly stated in one central document (e.g., a workplan, affiliation agreement, MOU, etc.); each organization has a formal and unique role in the collaboration; the organizations work together to develop new resources to support the collaboration; and communication among the partners is regular and highly developed. There will be many options beyond these two examples that partners will be able to identify and choose from within their own contexts.

Step 5.6: Select the Best Option

In order to select the best option among the approaches identified in Step 5, the participants must clarify what strengths and weaknesses each option offers. This analytic phase should begin with the development of characteristics on which the options can be compared.* To ensure that the key concerns and goals of stakeholders are reflected, a broad range of factors should be identified for each option, such as the following:

- Personnel
- Equipment
- Facilities
- Supplies/other
- Operational practices
- Sustainability
- Management support
- Constituency support
- Timeliness
- Relevance to consumer needs
- Quality of information

Next, the group should determine how to express their assessments of these characteristics. For example, personnel readiness could be rated as: (a) inadequate – significant resources needed above current levels; (b) marginal – limited resources needed above current levels; or (c) adequate – no additional resources are needed above current levels. The assessments need to be done for each organization that is included in each option. Differences in assessments and the uneven organizational impacts related to selecting each option should be fully discussed in the decision-making process.

Often, future benefits and costs of alternative approaches cannot be predicted with great confidence. Uncertainty may come from factors completely outside the policies of interest, outside of the partnering organization's control, and may affect the differences in impacts. The group should note how assumptions about future developments (e.g., changes in laws or elected officials) might drive success or failure of each option. This type of analysis should be conducted

* This analysis could be done on a structured, qualitative cost-benefit basis (see the Cost-Benefit Scorecard presented later in this chapter).

in an iterative process – group participants will need to change the information during the course of the analysis as new information and insights are gained. Although information from Phase 4 of the Framework will provide valuable inputs to this step, additional information may be needed here. Further, to ensure that everyone is aware of the quality of the information used in the inventories, a list should be provided of the kinds of uncertainties and assumptions made in completing these tools.

The choice of the best option must be sensitive to fiscal, political, and organizational realities and often involves many tradeoffs among the collaborators. These exchanges are most successful when the group has been able to sustain a climate of trust and respect throughout the Framework process. Typically no one option will beat all others. The use of the Scorecard or a similarly structured tool to visualize these tradeoffs, however, can facilitate discussion of the options. Choosing the optimal strategy may be complicated by the fact that people disagree about the assessments shown or the likelihood of different outcomes. These types of disagreements should be resolved within the group before a final "best" option is selected for action. When these disagreements cannot be readily addressed within the group, facilitation by a trusted community leader or hired facilitator should bring the decision to closure.

The final element of this step should be celebration! Getting this far into the Framework and having a specific option to pursue merits celebration of the group's success. It is valuable to take a moment to honor everyone's commitment to the process and to acknowledge the good that has been created by staying the course. Pausing to celebrate can also help motivate collaborators for the next phases – Prepare for Action and Take Action. The long-term benefits of the process are about to be realized.

Phase 6: Plan to Take Action

Because of their long-term natures, this and the next phase of the Framework were outside of the scope for the regional workshops. Nonetheless, some brief points will be made as these phases are essential components of an effective communication partnership. This phase focuses on getting a working plan and resources aligned so that the selected action can be taken. Careful planning and preparation will yield larger benefits than a more superficial effort.

Step 6.1: Plan the Process

A small group of representatives from key organizations may be the most suitable group for drafting an initial work plan to implement the option selected. Leadership of this group should be acceptable to all collaborators and a specific timeframe for completing a draft action plan should be clearly stated and shared with all partners. The planning group should also develop both short and long-term measurable objectives; these should be used to shape the activities and timeline that make up the action plan (below).

Step 6.2: Define the Action Plan

The first step toward accomplishing the option selected is drafting a comprehensive plan of action, comprised of a series of activities. For each activity identified, determine, for example:

- Which organizations will contribute to the activity
  - What services each partner will contribute
  - What resources will be provided by which partners
- Which organization will be responsible as the lead
  - Who will be the lead
  - How tasks will be coordinated across partnering organizations
- How each activity will be completed.
  - How tasks will be sequenced and linked

Step 6.3: Define the Measures of Success

The plan should also provide means to assess whether the actions taken meet the collaboration's goals. Key issues include:

- What measurable outcomes will be used as indicators of progress
  - How indicators will be fed back into the process for ongoing improvements
  - What process measures can be used
  - What outcome measures are feasible

All participants should then review the draft plan, provide their input, and accept a final action plan before it is implemented. It is imperative to have all partners fully informed about the plan, what it will require of their organization, and what to expect as results from its implementation. Any differences in expectations should be resolved as soon as they are recognized.

Step 6.4: Obtain Resources

Once the action plan has been prepared, the resources necessary for implementation must be secured. All resources do not have to be in place at the beginning of the plan, but they must be available by the time they will be needed in the timeline. Any problems in securing resources should be brought to the attention of the activity coordinator as soon as possible. The impacts of resource delays or failures will need to be considered by all partners, alternatives for accessing or leveraging other sources identified, or other strategic decisions made and actions taken. Limited resources may have serious consequences for the entire plan.

Phase 7: Take Action

The plan has been finalized and resources have been aligned when this phase begins. Problems that occur during this phase need to be identified as quickly as possible; often this is best done through the use of process measures (e.g., number of customer service calls), monitoring activities (such as obtaining a measure of customer satisfaction with education and outreach services), and/or indicators that identify trends (e.g., trends in types of questions raised by customers).

Step 7.1: Implement the Plan

Each activity lead takes action at the time indicated by the plan, or by the modified plan if prior changes have been made. Clearly, each lead needs to remain informed about the overall plan to conduct his/her work appropriately. When each activity is underway, the lead is responsible for ensuring that tasks are completed effectively and appropriately.

Step 7.2: Evaluate the Program

Finally, an often overlooked or under-resourced aspect of decision-making processes is evaluation of the tools and methods used. While process measures (such as the number of people who attend a public meeting, or the number of e-mail comments received, etc.) can offer useful measures, it is important to keep the ultimate outcome of the decision-making process in mind and develop measures for this outcome over time. The changes in key indicators over time (e.g., increased consumer support for infrastructure improvements to reduce and prevent health risks, reduced rates of specific adverse health outcomes, etc.) can provide important means for demonstrating positive returns on program investments. Outcome measures are often difficult, but not impossible, to design. Typically, elected officials and stakeholders engaged for a limited amount of time are more interested in process measures, while residents may want long-term outcome measures. A little bit of background work will likely reveal which people value which types of measures and why. Rarely will one type of measure meet all stakeholders' and managers' needs.

In this step, both process and outcome measures should be used to evaluate the team's progress in completing activities and moving toward the collaborators' vision. In the short-term, evaluation can focus on process measures and thereby provide a mechanism for continuous quality improvement. For example, the number of people who attend public meetings (e.g., relating to proposed water quality standards) could be used as a short-term measure of constituency support. Timeliness could be measured by determining how long it takes for information to be developed and distributed (e.g., CCR or public information brochure).

In the longer term, however, evaluation should increasingly focus on impact measures (e.g., changes in customers' knowledge about water-related health risks and interventions) and demonstrate ongoing value of the program of interest. The results from this type of evaluation will keep the partners committed to a successful collaboration and lead to the development of new and improved methods for collaboration. Customer surveys, focus groups, and/or website comments could be used to measure a communication strategy's relevance to consumer needs.

Step 7.3: Feed Findings Back Into the Program

Finally, continuous feedback loops – for staff, public, and other stakeholders – are invaluable mechanisms for obtaining insights as the basis for ongoing improvements in risk communication programs and strategies. These loops can be provided in meetings, on interactive websites, over telephone hotlines, through community groups, etc. There are a wide variety of methods that can be used to ensure sustained, interactive, and meaningful communication with stakeholders. When these options are established, it is important to acknowledge and demonstrate that the feedback is valued, seriously considered, and – as appropriate – leads to specific changes.

Evaluation is not useful if the results are not directed back into the collaboration's activities for continuous improvement. It is crucial to have routine ways of analyzing the evaluation results, interpreting them, and providing the findings to the relevant activity leads and decision-makers. Overall measures of progress are important to share with all partners to ensure that they are aware of both successes and problems with the plan. Occasionally, group discussions may be needed to brainstorm ways to address limitations and thereby reinforce the fundamental concept of collaboration: "we are all in this together." The sharing of the responsibility and successes is essential to sustaining the partnerships momentum until their vision is achieved.

Additionally, public sharing of the group's successes can bolster the team's enthusiasm as stakeholders recognize the impacts of the teams work. Getting public input may also lead to program improvements or mid-course modifications that strengthen the collaboration. Celebrating milestones achieved in the work plan can be used as opportunities to gain more public support for improved risk communication initiatives.

COST-BENEFIT SCORECARD

Once a jurisdiction has realized its vision, there will be many times when communication approaches will have to be evaluated and the best option chosen to address specific issues. The second tool, a Cost-Benefit Scorecard, focuses on assisting jurisdictions in both opening their thinking to more communication options and selecting the best method for a specific initiative. The analysis relies on many of the concepts used in the Framework, but while most cost benefit analyses are quantitative in nature, this tool utilizes a qualitative approach to balancing the options for action. The primary reasons for a qualitative approach are that many utilities do not have formal cost-benefit capabilities, and choosing among many risk communication options does not require such complex analysis or detailed inputs.

Analyses of the costs and benefits of implementing communication alternatives are important for determining the best way to allocate funds within a program. Given limited resources, decision makers and program implementers must determine not only what alternatives work best, but which ones do so at the lowest cost.

Cost-benefit analysis is a method for organizing information to perform this evaluation; it helps in separating acceptable from unacceptable alternatives, to ensure that resources are invested wisely (NAS 2004). In its most formal use, benefits and costs are expressed in monetary units, providing decision makers with a clear indication of the alternative that is the most economically efficient, or generates the largest net benefits (OMB 2003).

Characterizing impacts related to complex environmental health problems in monetary terms is often extremely difficult or impractical. Other frameworks are useful to consider when this is the case, or when objectives other than economic efficiency weigh heavily in decision-making. Here we present a qualitative framework for examining costs and benefits to help in choosing the best alternative for water-related health risk communication, building upon the RAND Corporation's policy scorecard analysis (Karoly et al. 2001). Our framework can help utility and public health agency personnel objectively consider multiple benefits and costs while drawing comparisons between alternatives, within a context of highly variable resources and other uncertainties that result when alternatives are applied to different settings and risk situations.

Table 4.3 presents the overall structure of the Cost-Benefit Scorecard and identifies the tools (presented at the end of this chapter) that link to specific steps.

Phase 1: Problem Definition

Step 1.1: Define the Problem

Analysts should begin by defining the problem they want to analyze and correct. For example, the goal might be to develop communication strategies and tools that would enable

Table 4.3
The structure and tools for the Cost-Benefit Scorecard

Phase	Step	Tool
1: Problem Definition	1.1. Define the Problem	C1.1
	1.2. Select the Options of Interest	C1.2.1 C1.2.2
2: Analysis	2.1. Develop Impact Measures	C2.1
	2.2. Develop Outcome Categories	C2.2
	2.3. Forecast Impacts	C2.3
	2.4. Document Assumptions and Uncertainty	C2.4
3: Preparation for Action	3.1. Select the "Optimal" Option	C3.1
	3.2. Evaluate the Program	C3.2

water utilities, the public health community, and clinicians to collaborate and thereby strengthen the utility's credibility with customers and their capabilities for addressing threats to drinking water quality.

In defining the problem, analysts should indicate who the decision makers are and who is affected by the problem, and note the extent to which there is agreement among these groups about the nature and importance of the problem. Then, they should determine the relevant time frame for analyzing solutions to the problem. The period chosen will depend on a number of factors, including the period of agency funding streams and the length of time and amount of resources needed to successfully address the problem. (See **Tool C1.1**)

Step 1.2: Select the Options of Interest

The analysts should begin this process by describing the baseline or the current state of the world without implementing a communication option. This is the scenario against which the risk communication options will be compared. The next step is to generate a set of communication options. Each option is represented in a column in a table (see **Tool C1.2.1**). Examples include:

- Baseline: Status Quo
- Option A: Leverage existing partnerships; could include developing informal links with community organizations, the medical society, or a university to expand communication services
- Option B: Develop legal tools for sustaining communication programs among public health agencies, utilities, and clinicians.
- Option C: Develop an ongoing water-health communication campaign in which a representative at the public health agency is the lead, but works in informal collaboration with clinicians and partners at the water utility.

The analysts should fill in the table with descriptive information about the baseline and alternative option scenarios (see **Tool C1.2.2**). In order to do this, they will need to answer questions such as:

- What is the *purpose* of each option?
- What is the *structure* of each option? Note roles of each party, linkages among parties, and where resources will come from.
- What *processes* will be used? Note the process for decision-making, leadership, and communication.
- What is the *time frame* to accomplish the communication program?

For example, in the baseline scenario, roles within the informal collaboration are loosely defined, no new resources are available for the collaboration, there is little or no joint decision-making, and only some informal communication occurs between the sectors. In collaboration option C, however, formal, written work assignments are in place; each sector has a formal role in the collaboration; the sectors work together to develop new resources to support the collaboration; and communication among the sectors is highly developed. See Appendix E for an example.

Phase 2: Analysis

Step 2.1: Develop Impact Measures

The analysis phase should begin with the development of impact measures, or measures of the costs and benefits of the communication options. To ensure that the concerns and goals of stakeholders are reflected, analysts should include a broad range of impact measures. These might include categories such as personnel, equipment, sustainability, timeliness, quality of information, and others (see **Tool C2.1**).

At this point in the process, analysts should also begin to develop measures of outcomes for evaluative purposes. In the short-term, evaluation can focus on process measures and thereby provide a continuous quality improvement function to the collaboration (Stokols et al. 2003). The results from this type of evaluation will keep the partners committed to a successful collaboration and lead to the development of new and improved methods for collaboration.

Step 2.2: Develop Outcome Categories

Next, determine how to express the impacts in qualitative outcome categories. For example, outcome categories for personnel needs could be classified as: (a) inadequate – significant resources needed above current levels; (b) marginal – limited resources needed above current levels; or (c) adequate – no additional resources are needed above current levels. (See **Tool C2.2.**)

Step 2.3: Forecast Impacts

Analysts should next develop a scorecard, and apply it to their circumstances. As shown in the example of **Tool C2.3** in Appendix E, the impacts of each scenario should be broken by stakeholder group and qualitatively scored using the impact measures and outcome categories developed in steps three and four above. In our illustrative scorecard, we include four stakeholder groups: the utility, public health agency, clinical entities (e.g., hospitals, other medical facilities, professional societies), and community organizations (e.g., neighborhood associations, local chapter of American Cancer Society, March of Dimes, etc.).

This scorecard demonstrates that costs and benefits may accrue to different stakeholder groups. These differences in impacts should be discussed in the decision-making process. In addition, analysts should consider how ongoing versus crisis situations might affect the performance of different programs.

Step 2.4: Document Assumptions and Uncertainty

Often, future benefits and costs cannot be predicted with great confidence. Where a range of categories is possible, the range should be explicitly stated in the scorecard (Karoly et al. 2001) (see **Tool C2.4**). Uncertainty may come from factors completely outside the policies of interest, and may affect the sizes of impacts. Analysts should make note of how assumptions about future developments may drive success or failure of the program (e.g., changes in laws or elected officials). In practice, this will be an iterative process – analysts will need to delete or add impacts during the course of analysis as new information is gained.

Phase 3: Preparation for Action

Traditional CBA converts costs and benefits into dollars, then notes which ratio of benefits to costs is greatest. In this qualitative analysis, however, analysts will need to interpret the costs and benefits described in the scorecard in a context relevant to the organizations involved by evaluating the complete set of options in terms of the complete set of goals/criteria. The choice of option must be sensitive to fiscal, political, and organizational realities.

Step 3.1: Select the "Optimal" Option

Typically no one option will beat all others on every impact measure, so selecting one program will involve trading one impact against another (Karoly et al. 2001). In the example table for **Tool C2.3** (see Appendix E), the option with the best "score" for each measure is shaded, while the worst is in italics. This can help in visualizing where the costs and benefits are located.

Choosing the "optimal" option is complicated by the fact that people may disagree about objectives or the likelihood of different objectives, or decision makers may value the impacts on certain stakeholders differently. These disagreements should be resolved among the decision makers, with the interaction of interested stakeholders. Decision makers must also consider how the options might work within their organizational structures, and how they might get the buy-in of their organizations.

Step 3.2: Evaluate the Program

Once the decision to implement a particular option is made, managers must be sure tools to evaluate the program are in place. Evaluation can provide a continuous quality improvement function to the risk communication collaboration (Stokols et al. 2003). It can also help analysts determine if messages are effectively reaching the public, demonstrate ongoing value of the program of interest, and keep the partners committed to a successful collaboration. Further, long-term evaluation can lead to greater recognition of the value of collaboration, institutionalization of the components, improved health interventions, and, ultimately, improved health outcomes (See **Tool C3.2**).

KEY POINTS

These tools to facilitate collaborations are based on the concepts that 1) complex issues are made easier through systematic approaches and 2) tools designed to assist participants through the process, with tangible and useful outputs, support participants through positive feedback and lasting benefits.

The **Framework For Action** was compiled based on "best practices" from the business literature and lessons learned in the data gathering components of the project.

- Values important to successful collaborations typically include trust, mutual respect, and honesty.
- Fundamental principles that often shape collaborations are systems thinking, strategic thinking, dialogue, shared vision, and shared success as well as responsibility and risks.
- Creating an effective collaboration is not an all or none proposition; it is a process that matures over time. Addressing each component of the Framework For Action assists the development of collaboration and fundamental relationships that influence the group's performance and outcomes.
- The tools and guidance that comprise the seven-phase Framework For Action are based on the development of mutual vision and values, and on increasing knowledge among partners about their roles and responsibilities. The use of the Framework leads to a series of tangible outcomes and opportunities for sustaining substantive collaboration over time.

The **Cost-Benefit Scorecard** was developed to guide the decision process for identifying the best option when choosing among a variety of possible communication approaches. It was based on cost-benefit literature as well as input we received during the project.

- The Scorecard aids in systematically organizing and analyzing complex information for determining efficiently which communication option will work most effectively at the lowest cost.
- Costs and benefits may accrue to different stakeholder groups, so differences in impacts should be taken into consideration when choosing the best communication option.
- Assumptions about future developments (such as changes in laws or elected officials) could drive the success or failure of a particular program; these assumptions should be explicitly addressed in developing a collaboration strategy.
- Program evaluation can help determine if messages are effectively reaching the public, demonstrate ongoing value of the programs, and keep the partners committed to a successful collaboration.

Last Updated by ______________
On (date) ___________________

FRAMEWORK FOR ACTION

Tool F1.1
Determine organizational need

Consider what problems exist and why. Here are some questions to stimulate and organize your thoughts. Share your draft need statement and bases with core participants in your organization. Get their input to prepare a one-page draft that can be shared with upper management for obtaining their support to initiate a collaboration planning process.

Question	Short answer	Supporting comments
1. Has there been a *recent event* (e.g., real or drilled emergency, emerging issue) that resulted in risk communication problems?		
What were the problems? (Specify)		
Which organizations were involved? (List)		
How important is it to address these problems with these organizations?		
2. Has there been a *change* that affects your organization?		
Does the utility have a new mandate or responsibility?		
Has there been a change in leadership?		
Has there been a change in organizational values, priorities, or concerns?		

(continued)

Tool F.1 (Continued)

Question	Short answer	Supporting comments
2. Has there been a *change* that affects your organization? (continued)		
Are there new societal contexts (e.g., public pressures, interests, etc.)?		
Are there new conditions that make this a time when innovative approaches are likely to be well received and implemented?		
3. Are the current issues *strategic* to your organization?		
Is addressing these problems crucial to your organization?		
Is addressing these problems crucial to the community?		
Will not addressing them affect the credibility or functionality of the organization?		
Will not addressing them now (or soon) result in larger scale problems that the organization will have to deal with later?		
4. What *need* does your organization have? State that need in a simple, short sentence.		
5. What are the most important reasons for this need? State them in a few, brief bullets.		

Last Updated by ______________
On (date) ___________________

TOOL F1.2
Identify potential partners

Organization:________________ Unit (if applicable):______________________ Sector:* _________________________

Contact Person: _________________ Phone:______________ E-mail: ______________ Other:______________________

Question	Short answer	Comments
1. How *familiar* are you with this organization/unit? Do others in your organization know this group? Does upper management in your group and this organization have a working relationship?		
2. What *roles* does this organization/unit have that relate to the need you have identified?		
3. What *services* does it provide that are related to the need? Are these services essential in meeting your identified need?		
4. Have you had *prior experiences* with this organization/unit? Have those experiences been largely positive? What problems have occurred? Is there a staff member in the organization who would be approachable about addressing problems?		

* e.g., Government (local, county, state, federal), health care providers (nurses, physicians, dieticians, etc.), education (schools, colleges, universities, libraries, etc.), first responders (emergency response, ambulance companies, fire, police), faith-based organizations (related to churches, temples, etc.), professional associations (medical society, teachers' organization, etc.), voluntary organizations (Red Cross, Lung Association, Rotary, Elks, Women League of Voters, etc.), business organizations (Chamber of Commerce, Better Business Bureau, etc.), philanthropic organizations (foundations), etc.

(continued)

TOOL F1.2 (Continued)

5. How *interested* is this organization likely to be in collaborating to address the need identified? What level of participation can be expected?		
6. What *concerns* would their staff have about collaborating? What barriers would they have to participating? How can they be addressed? Does this group have special needs that would have to be addressed? If so, what are they?		
7. What could this group *contribute* to a the process of developing a collaboration? What expertise or authorities do they have? Do they offer a unique or important perspective? Has this group or is it conducting a similar process you can build on? Can they contribute space, services, or other resources?		
8. What are the *defining characteristics* of this organization? Does it have power that could influence how the need is addressed? Is it recognized as a leader in the community? Does it represent important stakeholders? If so, which stakeholders? Is the organization making requests for your group's attention? Does it have priorities that align with your organization?		
9. Should this organization/unit be included in the *core* group to develop the Framework process?		
10. Is this organization/unit approachable at this time? What would be the most effective way to engage this group?		

TOOL F3.1
Develop a vision

Defining a vision involves developing a view of how things will work and by when the problem has been solved. The scale of the vision must relate to the scope of the problem. The visioning process should:

- Foster creativity,
- Be shaped by the participants, and
- Be driven by their priorities, issues, and desires to achieve improvements.

Answer open-ended questions (like these examples) in broad terms, not in detail:

- What does good risk communication mean to you?
- What are the key characteristics of good risk communication?
- What risk communication program do you think the community should have five years from now?
- How would that program work? Who would be part of it?
- What would the impacts of good risk communication be?
- What will be the benefits when your vision of risk communication is achieved?
- What will happen if your vision is not achieved?

The group's responses to these questions should result in a brief statement of the group's ultimate goal for the Framework process. It should capture how the participants want to see risk communication occurring at a defined point in the future. The statement should be inspirational, motivational, and displayed frequently.

TOOL F3.2
Agree on values

Values are the principles and beliefs that define and guide the group process. They set out how the participants want to interact, and what they expect of each other and the process. The group's shared values provide the essential foundation for the entire Framework process.

Answer open-ended questions (like these examples) in broad terms, not in detail:

- How do you and your organization want to be treated?
- What do you expect of the group?
- What behaviors do you want the group to demonstrate?
- What do you think it means to be a team?
- What do you think it means to collaborate?
- What kind of working environment do you want the group to have?
- What do you believe the group needs to achieve?
- What would help you stay involved, engaged, dedicated to this process?

At the conclusion of the values discovery process, the values should be written in a brief, simple statement that is accepted by all participants.

Last Updated by____________________
On (date) __________________________

TOOL F3.3
Define types of information needed

Categories of information need to relate to the vision, problem, and contexts of the problem. The objective of this step is to identify the types of information and sources of those types that will be used in comprehensive assessments. Participants must be fully informed, understand, and then learn from current conditions before they can identify what the underlying causes of their risk communication problems are. The purpose of this step is to ensure that all participants are informed about the types of information they will need to gather to implement the Assessment phase.

Check off the types of information that are available within and external to each participant's organization. Note how accessible this type of information is (e.g., easy, difficult, not accessible, unknown), how current the information is, and who is responsible for the information.

Type of Information	**Available**	**Accessible**	**Most recent information**	**Contact person**	**Contact information**
Internal to your organization					
Strategic issues Board of Directors' priorities					
Types of services provided Risk communication materials Current collaborators and their contributions Stakeholders of these services Other stakeholders who need services					
Evaluation of services (e.g., timeliness, relevance, quality, effectiveness, etc.) Risk communication materials Current collaborators and their contributions Stakeholders of these services Other stakeholders who need services					

(continued)

Tool F3.3 (Continued)

Type of information	Available	Accessible	Most recent information	Contact person	Contact information
Resources dedicated to risk communication services Personnel Equipment Office space Supplies/other support Operational policies and practices Budget Management support Sustainability					
External to your organization	**List short answers**				
Others' needs for your services					
Others' expectations of your services Assessments of your services					
Others' support for your services					
Others' willingness to help improve your services					
Population trends Growth rate Diversity Ability to use computers Knowledge of drinking water issues Knowledge of public health issues Sources of relevant information Credibility of your organization as a source of drinking water-related risk information Trust in government					
Other trends (e.g.) Societal Economic Political Legal					

Last Updated by____________________
On (date) ___________________________

TOOL F4.1.1
Inventory of water-health risk communication materials and services

Answer the questions for each material or service listed. If your organization has other relevant items, list them at the end of the table and answer the same questions.

Risk communication materials	**Is it available?**	**Where is it kept?**	**Who is the lead?**	**When was it last updated?**	**Which partners provided input?**	**What is its primary purpose or role?**
Written tools						
Brochures (on specific contaminants or more general health-related topics)						
Consumer Confidence Reports						
Door hangers (e.g., boil water alerts)						
Draft press releases on health issues						
Emergency response plans						
Fact sheets on water-health issues						
Public notifications						
[add others]						

(continued)

TOOL F4.1.1 (Continued)

Risk communication materials	Is it available?	Where is it kept?	Who is the lead?	When was it last updated?	Which partners provided input?	What is its primary purpose or role?
Visual tools						
PowerPoint presentations						
Slide presentations						
Overheads						
[add others]						
Computer-related tools						
E-mail service for questions from the public						
Website						
[add others]						
In-person services						
Clinical outreach (e.g., Grand Rounds)						
Programs open to the public (tours, seminars, etc.)						
School-based Programs						
Speakers bureau						
[add others]						
ADD OTHER SERVICES OR MATERIALS						

Last Updated by___________________
On (date) _________________________

TOOL F4.1.2
Identify current partners

List the organizations that you partner with for the purposes of drinking water-related health risk communication. For each partner, answer the questions to the best of your ability.

What is the name of your partner organization? [insert under categories, such as those noted]	**Who is your primary contact there?**	**How often do you and the lead contact communicate?**	**What services does the partner provide to your organization?** [e.g., regulatory, media outreach, etc.]	**How critical are these services to your programs?**	**What factors contributed to starting the partnership?**	**When did the organizational partnership begin?**
Utilities						
Local public health agencies						
Clinical care organizations (e.g., hospitals, clinics, dialysis centers, etc.)						

(continued)

TOOL F4.1.2 (Continued)

What is the name of your partner organization? [insert under categories, such as those noted]	**Who is your primary contact there?**	**How often do you and the lead contact communicate?**	**What services does the partner provide to your organization?** [e.g., regulatory, media outreach, etc.]	**How critical are these services to your programs?**	**What factors contributed to starting the partnership?**	**When did the organizational partnership begin?**
Local emergency response organizations						
Police departments						
Fire departments						
Other local agencies						
Voluntary organizations						
Community associations						

Last Updated by____________________
On (date) ________________________

TOOL F4.2.1
Identify mandates and authorities

Communication programs may be legally required, but these legal bases may be poorly understood by staff who implement the programs. The purposes of this inventory are to (1) bring the foundations for risk communication to a higher level of staff awareness and thereby (2) strengthen appreciation for the significance of mandates and authorities in relationship to organizational missions and the collaboration's foundations. Complete the table as much as possible for your own organization.

Governmental level and type of legal instrument	What is the name of the legal instrument that mandates related roles and/or services?	What are the relevant section/s of this instrument?	What organizational roles and/or service/s are required?	What responsibilities do staff have under this mandate?	On what date did the mandate take effect?
Local					
Ordinance					
Other					
County					
Regulation					
Other					
State					
Law					
Regulation					
Other					
Federal					
Law					
Regulation					
Other					

TOOL F4.2.2
Assess programmatic resources

The type and level of resources required to support and sustain water-related health risk communication services need to be evaluated qualitatively. Exact data or audited statements are not necessary to address the following questions.

For each service,

- Are resources sufficient to support current services?
- Are resources expected to be sufficient to continue these services?
- Are some services or materials no longer needed and therefore ripe for discontinuation?
- Can resources used for duplicative approaches be redirected into complementary or missing services, or are the resources "blocked" for these services only?

TOOL F4.2.3
Assess past performance and emerging trends

Partners can gain insights from looking both backward and forward. This tool is designed to assist collaborators in assessing the past and considering trends that may influence the future.

To assess past performance, consider questions such as:

- What have been the impacts of the services provided?
- Have the services performed as expected?
- Are the services valued by the end-users?
  - Are people satisfied with them? Do they find them easy to access, use, etc.?
- Are the services making enough of a difference to continue them as they are, or should they be modified?

Routinely conducted customer surveys, telephone logs, or other ongoing data collection may offer insights about past performance and trends in customer satisfaction and impacts.

To identify trends that may affect your organization's risk communication services, consider questions such as:

- Are there demographic trends that are reshaping important characteristics of the population served?
  - Will services have to change to meet the new population's needs? In what way?
  - Are community groups already reaching these populations? Can we partner with those groups?
- Are economic or technological trends creating new demands for or concerns about risk communication services?
- Are political trends likely to result in changes in routine or emergency communication methods?
- What societal trends are likely to influence risk communication strategies?

- Are public or professional perspectives on water-related health risks changing?
  - How important are these trends in people's decisions related to water-health issues?
  - How will these changes affect the ability of stakeholders to receive and act on the information as organizations intend?
- Could any of these trends influence the nature and value of risk communication activities as they are now conducted?

TOOL F5.1
Identify strategic issues

Strategic issues are fundamental choices or challenges that *must be addressed* for the group's shared vision to be realized. They are more than critical in nature; they elicit proactive thinking, positioning the collaboration toward the future, and seizing current and emerging opportunities to make progress toward the collaboration's vision. The following questions are examples to guide the recognition of strategic issues.

Based on the data and information shared to date:

- Which problems contribute to other issues?
- Are there additional problems that need to be added to the list?
- How do these new problems relate to the ones we already identified?
- Which problems are the *major* issues?
- How do these issues relate to themes we have identified? How do these issues relate to each other?
- What would happen if this problem is not addressed? Would the consequences limit the group's ability to achieve its vision?

TOOL F5.2
Recognize why issues are strategic

Discussion about how the identified problems relate to participants' organizational mandates, authorities, resources, personnel, operations, etc. should help the group understand the complexity and significance of the problems. When the participants recognize how a problem fits with their own organizations, it may then be easier for them to discuss how it fits with the group as a whole. Discovering efficiently which issues are fundamental to the collaboration may require the assistance of a neutral, skilled facilitator.

For each issue, consider the following (for example):

- Does it relate to organizational mandates and/or authorities?
- Does it significantly affect organizational resources, staff capacity, operations, service quality, or performance?
- Is it irrelevant or weakly related to the group's vision?
- Is it urgent?
- Is it affected by external threats or trends?
- Is it related to internal weaknesses or opportunities?

- Will there be negative consequences if this problem is not addressed? What will these impacts be? How seriously would they affect the group's abilities to achieve its vision?
- Is it fundamental to the collaboration?
- *Must it be addressed* to reach the group's shared vision?

After all issues have been assessed, determine which issues are most fundamental to the participating organizations and which are most directly related to the collaboration's focus.

TOOL F5.3
Sequence the strategic issues

The list of strategic issues must be reduced to about ten specific statements that characterize the fundamental problems related to the shared vision. Statements of the strategic issues must not overlap or omit significant problems. In the process of finalizing the list of the major strategic issues, new concerns or new ways of framing the issues may be discovered. The first part of developing an order for the issues, however, is the development of a short statement of each major strategic issue.

Each strategic issue in the final list must then be evaluated for its impact on other issues and in larger contexts. Questions to consider include:

- Will solving this issue mean that other issues will also be addressed, at least in part?
- Does resolving this issue require solutions to other issues, in whole or in part?
- Are there any unique opportunities (e.g., events, trends, etc.) that we could build on to address this issue?
- Is it the "right time" to address this issue now? Does it have urgent components? Does it fit with current trends or near-term events?
- Should this issue be addressed – in whole or in part – at a specific time (e.g., linked to an annual event or process)?
- Is this an issue that should be addressed later or in the long-term?
- Does this issue require a series of actions that must be spread out over time (e.g., due to resource limitations, needs for others' input, anticipated regulations, etc.)?

Only in the full societal, political, economic, and other contexts can issue ordering be considered. As a result, the final list of when issues need to be addressed will be unique for each partnership.

TOOL F5.4
Define the goals

Bridging what *is* to what the group *wants* requires clearly stated goals.

Goals are broad statements of what outcomes the group wants to achieve in the long-term by addressing the strategic issues.

When the goals have been reached, and thus all of the strategic issues addressed, the group's shared vision will have been realized.

A few – typically about four – comprehensive, but not overlapping, goal statements should clearly state ***what outcomes*** are necessary for the group's vision to be accomplished. Goals need to be explicitly linked to the vision and strategic issues. Goals should clarify what must be done to address each part of the vision and with each strategic issue.

Goal statements can be developed by responding to questions such as the following:

- What are the components of our vision?
- What outcomes must we have to realize each part of our vision?
- What outcomes are necessary to fully accomplish our vision?
- What results will we have to have in order to address our strategic issues?
- How do the outcomes we want relate to each other?

TOOL F5.5
Develop the options

To accomplish the goals, the collaborators must determine ***how*** they will achieve them; this should be done by exploring a range of strategic alternatives. This step should not be bounded by what *is*. To develop new approaches, "thinking outside of the box" and collective learning that produces creative strategies is essential.

The final list of options should include no fewer than three and up to five clearly different alternatives. These approaches could be the status quo (no change), the ideal (requiring the greatest change), and one to three options in-between the two extremes. It is important to draft and consider comprehensively a wide range of possible approaches to achieving the vision.

Once the list of options has been developed, for each option consider questions such as:

- What is the *purpose* of this option?
  - Is it clearly linked to the vision and goals?
- What is the *structure* of this option?
  - What are the roles of each organization?
  - What are the linkages among the organizations?
  - What are the sources of all of necessary resources?
  - What forms of leadership and partnership would be required?
- What *processes* will be needed?
  - What processes for decision-making would be used?
  - What type/s of communication among the partners would be needed?
- How much *time* will it take to accomplish this option?
- What *resources* will be needed?
  - Are the resources available? If not, are they obtainable?
- Is this option *feasible*?
  - Will new legal authorities be required?
  - Will major policy changes be necessary?
- Will this option be *acceptable* to stakeholders?
  - Will the cooperation of key stakeholders be essential?
  - Will a range of stakeholders be willing to support this option?
- What is the *likelihood* that this option will be successfully completed?

Because local conditions vary widely, no two partnerships are likely to develop the same set of options.

Last Updated by___________________
On (date) ___________________

COST-BENEFIT SCORECARD

TOOL C1.1
Define the problem

The objective of this step is to develop a clear, mutually accepted statement of the risk communication problem. The questions below can be used to organize your thoughts and stimulate discussion.

Question	Answer
Who are the decision makers?	
Who is affected by the problem?	
To what extent is there agreement among these groups about the nature and extent of the problem?	
What is the relevant time frame for analyzing solutions to the problem?	

Last Updated by___________________
On (date) ___________________

TOOL C1.2.1
Select the options of interest

The collaborators need to describe the baseline or current state of the world as the status quo, the alternative in which no changes are made, as well as a set of potential options, which should range from ones that require minimal change to ones that entail extensive, new policies, programs, and partnerships. One option should describe the ideal option, regardless of whether it requires extensive change or not. In this step, it is important to consider a wide range of possible approaches for addressing the problem. See Appendix E for examples.

Option	**Description**
Baseline (status quo)	
Option A (minimal change)	
Option B (moderate change)	
Option C (ideal)	

Last Updated by____________________
On (date) ____________________

TOOL C1.2.2
Key characteristics of each option

Once the list of options has been developed, collaborators should consider questions such as:

- What is the *purpose* of this option?
- What is the *structure* of this option? (Note roles of each organization, linkages among the organizations, and the sources for all necessary resources.)
- What *processes* will be used? (Note the processes for decision-making, leadership, and communication.)
- What is the *time frame* to accomplish the option?
- What *resources* will be needed?

The following chart can help in visualizing the similarities and differences between options. See Appendix E for examples.

	Baseline	**A**	**B**	**C**
Purpose				
Structure				
Roles/Structure				
Links				
Resources				
Process				
Decision making				
Leadership				
Communication				
Time				

Last Updated by___________________
On (date) ___________________

TOOL C2.1
Develop impact measures

In order to select the best of the communication options, the participants must develop characteristics on which the options can be compared. To ensure that the key concerns and goals of stakeholders are reflected, a broad range of factors should be listed, such as the following:

- Personnel
- Equipment
- Facilities
- Supplies/other
- Operational practices
- Sustainability
- Management support
- Constituency support
- Timeliness
- Relevance to consumer needs
- Quality of information

Last Updated by____________________
On (date) ____________________

TOOL C2.2
Develop outcome categories

Next, the group should determine how to express their assessments of these characteristics. See an example table in Appendix E for more information.

	Outcome category 1	**Outcome category 2**	**Outcome category 3**
Impact measure 1			
Impact measure 2			
Impact measure 3			
Impact measure 3			

Last Updated by____________________
On (date) ____________________

TOOL C2.3
Forecast impacts

The goal of this step is to break out the impacts of each scenario by stakeholder group and assign a qualitative score, using the impact measures and outcome categories developed in earlier steps. This scorecard will help identify differences in impacts among the scenarios to the stakeholder groups. An example of this scorecard is given in Appendix E.

	Baseline	A	B	C
Stakeholder group 1				
Impact measure 1				
Impact measure 2				
Impact measure 3				
Impact measure 4				
Impact measure 5				
Stakeholder group 2				
Impact measure 1				
Impact measure 2				
Impact measure 3				
Impact measure 4				
Impact measure 5				
Stakeholder group 3				
Impact measure 1				
Impact measure 2				
Impact measure 3				
Impact measure 4				
Impact measure 5				

Last Updated by____________________
On (date) ____________________

TOOL C2.4
Document assumptions and uncertainty

The goal of this step is to discuss how assumptions about future developments might drive success or failure of each option. To ensure that everyone is aware of the quality of the information used in the scorecard and the related conclusions, a list should be provided of the kinds of uncertainties and assumptions made in completing the scorecard along with the scorecard outputs. See Appendix E for some examples.

Uncertainties

•
•
•

Assumptions

•
•
•
•
•

TOOL C3.1
Select the "optimal option"

In this step, the goal is to use the scorecard to visualize the tradeoffs and stimulate discussion of the best option.

Choosing the "optimal" strategy may be complicated by the fact that people disagree about the assessments shown or the likelihood of different outcomes. These types of disagreements should be resolved within the group before a final "best" option is selected for action. A trusted facilitator may need to aid the group.

Last Updated by____________________
On (date) ____________________

TOOL C3.2
Evaluate the program

The changes in key indicators over time can provide important means for demonstrating positive returns on risk communication program investments. In this step, participants should begin developing a list of key indicators, as well as evaluation measures for both the short and long-term. Some examples are given in Appendix E.

Key indicators
1.
2.
3.
Process measures (short-term)
1.
2.
3.
Impact measures (longer term)
1.
2.
3.

CHAPTER 5
RECOMMENDATIONS

INTRODUCTION

The participants in this project indicated that the general public and some specific subpopulations, that may be at higher risk than the general population for adverse health outcomes (such as HIV/AIDS patients, pregnant women, chemotherapy and dialysis patients, etc.), have voiced increasing concern about how their drinking water affects health. Their questions need to be answered with credible and timely information.

The research conducted in this project demonstrated that utility, public health agency, and clinical personnel want to implement more effective ways of addressing the public's drinking water-related health risk issues. The public good can be more effectively served through substantive collaborations in which these professionals share their knowledge, experience and perspectives. The challenge in most locations has been bringing representatives of these three sectors together and sustaining their relationships. The focus of this project was to create new means by which these and other sectors could develop more meaningful partnerships for addressing emergency, ongoing, and future water-health issues.

It is evident that there is no "one-size-fits-all" method that will fit the conditions in all localities. Instead, each jurisdiction needs to build on the lessons learned by others, and design an approach suited to the resources and commitments available in their area. The two toolkits presented in this report – the Framework For Action and the Cost-Benefit Scorecard – are offered as structured processes to help utilities and their partners organize and assess complex information. Using these toolboxes, they will be able to initiate crucial dialogues, make more informed decisions, and sustain relationships for more effective risk communication initiatives.

Many of the findings in this report point to recommendations that apply to all potential partners. These will be discussed before sector-specific recommendations are noted. However, each sector is encouraged to examine the recommendations for other sectors so they will build their awareness of the challenges and opportunities their potential partners face.

FOR ALL ORGANIZATIONS AND INDIVIDUALS

- Recognize that successful collaborations rely on respectful relationships, honesty, trust, and dialogue. These characteristics are dynamic, requiring continuous attention and nurturing. Whenever problems or misunderstandings arise, they should be treated as top priorities and addressed promptly.
- Question assumptions about each other's roles, responsibilities, resources, authorities, integrity, etc. Whenever partners find that they are making assumptions about each other, sharing of information and facts will reduce the likelihood of inaccurate guesses. Assumptions can seriously damage relationships, resulting in the need for commitments to repair in order for the collaboration to be sustained. Ignoring such damage will weaken partnerships.
- Cultivate effective relationships *before* an event (such as an emergency response or disease outbreak) occurs. Applying the Framework or a similar process proactively is

essential to position the partners for action under stressful conditions; practice during routine operations will provide valuable experience to build on in an event.

- Develop a sound process and structure to increase the probability that collaboration will not only be established but also be successful and sustainable. Further, the more complex the contexts (e.g., metropolitan areas with multiple jurisdictions as compared to a small or less complex service area), the more important formal processes and instruments become. As the number and diversity of partners increase, these tools provide more efficient means for communication and development of shared concepts and principles.
- Understand that collective information produces much richer knowledge about communication contexts than any one organization could assemble alone. In an effective collaboration, each partner contributes something unique – a perspective on drinking water quality and technology, public health issues, population risk perceptions, access to key community or other leaders, resources, etc. The sum of their contributions will be greater than the collection of the parts; together their knowledge provides a more comprehensive view of conditions and more readily reveals major themes and trends.
- For more creative tasks such as brainstorming a shared vision and values, use a neutral, skilled facilitator so all participants can fully engage in the dialogue.
- Recognize that collaborations emerge from an ongoing process; they do not occur from a single meeting or written document. Partnerships are dynamic; they change over time as their purposes, activities, and participants change, but the fundamental values and vision of the collaboration should remain stable for extended periods (e.g., five to ten years) and serve as the common ground in the face of change.
- Commit to some level of regular, in-person dialogues. Effective dialogue is essential for collaborations. Although e-mails and websites can be used for some communications, collaboration depends on rich exchanges of information and views that are most effectively completed in person. There is no substitute for face-time when building partnerships.

FOR THE PARTICIPATING UTILITIES

First, we would like to thank the five utilities that participated in the workshop and regional exercises, and the many utility staff members who contributed to other parts of this project. Each of them shared their successes and frustrations in establishing water-health related partnerships. The recommendations we provide are based on the input of the particular utilities and are offered to help them continue in their efforts to build effective collaborations.

- Continue to explore your approach to identifying partners, so you will not miss a potentially important supporter for your efforts.
- If you serve several or many jurisdictions, determine who their strategic stakeholders are; i.e., using the method provided in Chapter 3. Focus on these areas to establish collaboration and expand as needs dictate. There may be no need to engage all jurisdictions, but a careful analysis of stakeholders' characteristics may be pivotal to determining how to start or strengthen an existing partnership in complex contexts.

- Set up an opportunity for all partners to exchange additional information about their organizations' roles, responsibilities, and mandates. Using tools in the Framework as means for common ground, assumptions about each other can be reduced and new opportunities for progress identified.
- Review and improve communications between the agencies' and organizations' Public Information Officers. In some cases, PIOs' knowledge and abilities to work on water-health issues is sufficient, but more often confusion about PIO roles and responsibilities have created barriers for both routine and emergency communications. Issues that need to be addressed include which (if any) PIO would be the lead for water-health issues or types of these issues, who should craft messages and methods for the partnership, and how should PIOs work together.
- Address concerns about the lack of debriefings after emergency responses and drills. One approach would be for the partners to hire a neutral, skilled facilitator to ensure that important lessons are extracted from these events and converted into constructive appropriate policy and/or procedural changes.
- Evaluate risk communication materials and methods. Few written materials have been pretested among representatives of the target audiences.
- Evaluate the effectiveness of standard methods of communicating with clinicians, like broadcast fax approaches. While there is widespread awareness that broadcast fax contact lists are incomplete, there is no understanding about how many people on those lists actually receive the faxes at all or in a timely manner. Assuming that S/LPHAs can reach all clinicians is an important misconception.

FOR THE WATER UTILITY INDUSTRY

- Recognize that health care providers include more than physicians. In some areas, dieticians, nurses, dentists, pharmacists or other clinicians may be important partners to enlist in water-health collaboration. Efforts should be made by the industry to help utility personnel think more broadly about "clinicians."
- Work beyond current boundaries. Partnerships for water-health concerns may not be best limited to utility, public health, and clinical representatives. In many areas, community or voluntary organizations, school or college employees, emergency response, and other organizations may be essential to a water-health collaboration. As for "clinicians," the term "partner" needs to be broadened. One way to work past current boundaries is to encourage utilities to apply the stakeholder paradigm provided in Chapter 3. Seminars and workshops at regional or national meetings could build industry-wide skills in applying, interpreting, and using the results of a systematic stakeholder analysis.
- Strengthen representation beyond the governmental sector. Public health services are provided by more than local or state governmental agencies. Community clinics, voluntary associations, academic centers, hospitals, private laboratories, or other organizations play important roles in the public health network. Partnerships that address public health issues may be strengthened by representation beyond the governmental sector.
- Explore ways to assist S/LPHAs in building capacity and/or obtain resources to serve as the utility-clinical bridge they expect. Utilities consistently value public health

agencies as evaluators and interpreters of public health data and information, and representatives and/or links to clinicians. However, they often do not realize that S/LPHAs have limited capacity to meet utilities' requests, including limited capacity to contact clinicians.
- Apply and evaluate the Framework and Scorecard as a follow-up research project to this one, which was focused on developing the evidential base for these tools.
- Evaluate the key factors that foster and hinder success to provide important insights and guideposts for future collaborations.

FOR THE PUBLIC HEALTH COMMUNITY

- Develop resources to support the function of serving as the bridge between utilities and clinicians. Innovative ways to meet these sectors' expectations need to be explored. Agencies may need to seek partners to help them meet the needs of utilities and clinicians; professional associations and public libraries may be able to help with some education and information functions.
- Inventory the existing materials and programs that relate to drinking water. Often, divisions within public health agencies – especially large ones – may not be aware of each others' water-related health information and services. Agencies, across units – e.g., responsible for HIV/AIDS populations, maternal and child health, environmental health, health education, cancer registry, infectious diseases, elderly and other populations – may all have or need water-health knowledge and materials.
- Explore new strategies to reach clinicians, particularly those who serve populations at increased risk for water-related health risks; strategies may include leveraging existing resources. In many cases, these clinicians include specialty physicians and nurses, dieticians, directors of dialysis centers, etc. Options may include: setting up a website where relevant clinicians could enter and update their contact information, evaluation of existing methods for completeness and timeliness, seeking resources from government or a major utility or other sources, etc.
- Help the public and clinicians locate credible sources of information about water-related health issues. Fulfilling this role demonstrates that agencies are willing to provide their unique expertise to address these needs. Areas that do not now provide this service may need to locate resources to support it; some public libraries may be willing to help with this service.
- Integrate drinking water and health issues more extensively into existing programs; e.g., information provided to HIV/AIDS patients, cancer and dialysis patients, and other at-risk populations already served by the agency. Often this expansion will require a small investment compared to the health risks prevented.

FOR CLINICIANS

- If you serve at-risk populations, ensure that you have up-to-date knowledge about the relationship between water hazards and health. Utility and public health personnel expect clinicians to provide accurate information to patients about drinking water and health. When this is not possible, these sectors expect clinicians to contact credible sources for this information. Clinicians serving at-risk populations need to ensure that they have up-to-date knowledge about the relationship between water hazards and health.

- Communicate with your local public health agency through reporting illnesses that may be water-related and by calling the agency with questions. Building rapport in advance of an outbreak results in proactive prevention and makes the S/LPHA aware of your interest in your patients' water-related concerns.
- Recognize drinking water utilities as additional resources to answer patient questions on drinking water and health. Many utilities have a director of media or public relations who is skilled at routing clinicians' questions to the most appropriate staff member.

It is clear that many actions can be taken to establish and strengthen collaborations to address drinking water-health issues. The tools and recommendation offered in this report offer many options for partners to make progress in their partnerships. These are by no means the only possible methods to consider. We challenge our readers to start from where you are, find the tools that will serve your purposes, and identify interested partners.

APPENDIX A
PHILADELPHIA WATER QUALITY COMMITTEE MODEL*

* From Dr. Caroline Johnson's slide presentation, "Building a Collaboration Between the Water and Health Departments in Philadelphia" in March 2004.

SYNOPSIS OF DR. JOHNSON'S SLIDE PRESENTATION DURING THE MARCH 2004 WORKSHOP

In 1995, the public water supplier for the City of Philadelphia, Pennsylvania, began monitoring *Cryptosporidium* levels in drinking water. However, the utility did not contact the Philadelphia Department of Public Health (PDPH) about this initiative, had not estimated potential public concern accurately, and had not identified a spokesperson should public concerns arise. At the same time, the PDPH did not have health surveillance activities in place, had no knowledge of water processes, and had no mechanisms to communicate with health care providers or susceptible subpopulations (such as HIV/AIDS patients). After the utility's monitoring activity became known, PDPH's AIDS program staff notified PDPH's Division of Disease Control personnel that they were aware of 50 cryptosporidiosis cases per year.

Learning from the problems that arose from these circumstances, steps were taken to establish a collaborative partnership for effectively addressing future drinking water-related health issues. Initially, many stakeholders came to the table (Figure A.1). Over time partners dropped out (Figures A.2 and A.3) as the needs, purposes of collaboration, and roles of stakeholders became clearer. The community and special interest groups had different interests from the other stakeholders, and ultimately set up a separate group, the Water Quality Advisory Group, to meet their particular needs.

A formal agreement was reached between the utility and PDPH that provided the bases for the following activities:

- Water data is transmitted to PDPH on a regular basis
- Consumers are engaged in discussions about water issues early and often
- Spokespersons are identified before events occur
- Public communication messages to be used during events are drafted in advance
- Cryptosporidiosis is a disease reportable to PDPH
- Active surveillance for cryptosporidiosis is conducted by PDPH
- Survey data are sent to the utility by PDPH
- PDPH has a rapid, wide communication system utilizing telephone and broadcast fax technology for reaching health care providers
- PDPH and the water utility meet monthly to review data, and make technical decisions
- PDPH and the utility also determine whether the threshold for risk communication has been reached and – if so – what level of communication (in the tiered communication process) is appropriate to use

The overarching lesson learned was: Build collaboration BEFORE a stressful event occurs. Trying to build collaboration during an event is not ideal.

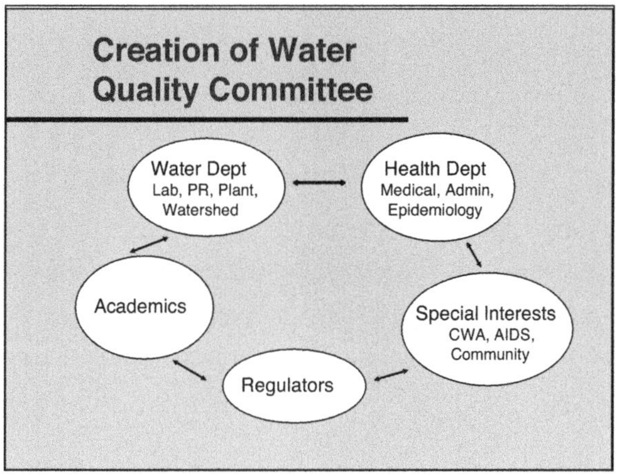

Figure A.1 Initial set of collaborators

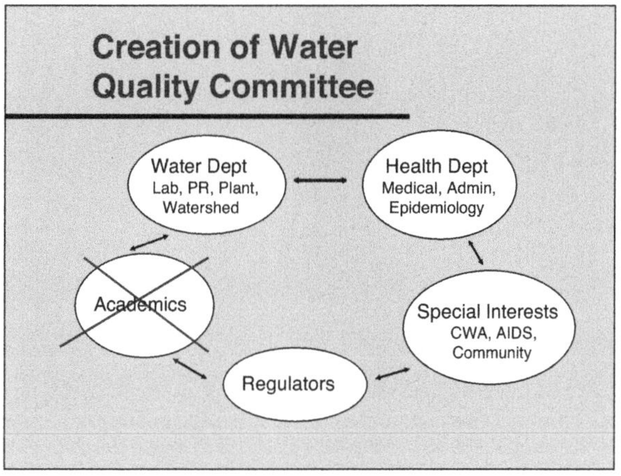

Figure A.2 Intermediate set of collaborators

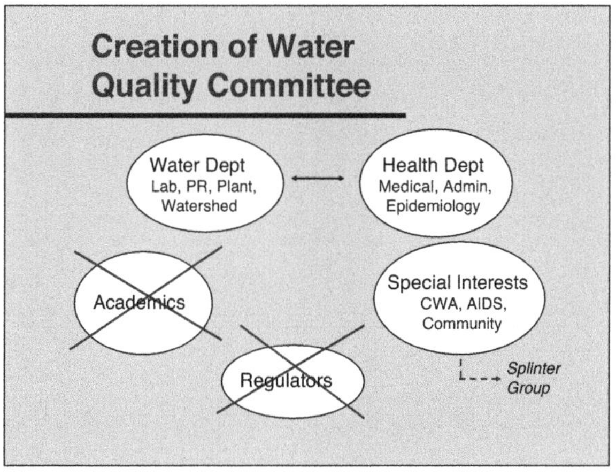

Figure A.3 Final set of collaborators

APPENDIX B
SURVEY AND INTERVIEW INSTRUMENTS

SURVEY QUESTIONS FOR WATER UTILITY PERSONNEL

Introduction

Good morning/afternoon. I am (name) … from Opinion Search, calling on behalf of The George Washington University. I am calling in follow up to our recent letter about our survey on communication practices of drinking water utilities.

Is this a good time to go through the questions?

[NO]
May I schedule a time to call you again?

Appointment (Day, Date, Time) ___________________________________

[Yes]
Let's continue.

All the information you provide us will be kept strictly confidential. Only the project staff will have access to your responses with no personal information attached. No one else will have access to this information without your knowledge. The Institutional Review Board of The George Washington University Medical Center has reviewed and approved this questionnaire and the study procedures.

May I continue with the questions?

[If the participant wants to talk with the study's participant representative, s/he may call the GWUMC IRB at 202-994-2764. In that case, ask if you may call the participant back tomorrow or the next day.]

Q1. Gender [DO NOT ASK. Check off the answer.]
1. Male
2. Female

Thank you for agreeing to participate.

First, I would like to ask you about how your drinking water utility operates and works with government agencies.

Governmental Structure and Organization

Q2. Which agency or agencies regulate drinking water in your state? [read, check all that apply]

		Yes	No	Don't Know
1	The Health Department or Agency			
2	The Department of Environmental Quality (DEQ) or Natural Resources (DNR)			
3	The Energy Department			
4	A Public Utility Commission, or			
5	[Do not read] Other [Specify]			

Q3. Is your utility considered [read]
1. Private
2. Public, or
3. Both

Q4. Is your utility governed by one or more of the following? [read, check all that apply]

		Yes	No	Don't Know
1	An Appointed Board			
2	An Elected Board			
2	Public elected officials, or			
3	Public appointed officials			
4	[Do not read] Other [Specify]			

Q5. Does your utility have a citizen or consumer advisory board?

1. Yes

0. No [Skip to Q7]

8. Don't know [Skip to Q7]

Q6. Does the citizen advisory board include representatives from any of the following sectors? [read, check all that apply]

	Group	Yes	No	Don't Know
1	Emergency Responders			
2	Health Care Providers, or			
3	Health Officials			
4	[Do not read] Other			

Utility Communication With Public Health Officials

Now I would like to ask you about how your utility and the public health departments or agencies communicate in your area.

[Read if necessary: Communication refers to both formal and informal contacts where information is exchanged]

Q7. Does your utility work with: [read, check all that apply]

		Yes	No	Don't Know
1	Local Public Health Agencies			
2	State Public Health Agencies			

Q8. Which of the following best describes how often your utility communicates with health agencies? [read]

1. Daily
2. Weekly
3. Monthly
4. Quarterly
5. Yearly or less
6. Never [Skip to Q13]
7. [Do not read] Other

Q9. When your utility communicates with health agencies, which of the following methods are used: [read, check all that apply]

		Yes	No	Don't Know
1	Conferences			
2	E-mail			
3	Fax			
4	Mail			
5	Meetings, or			
6	Phone			
7	[Do not read] Other [Specify]			

Q10. Which best describes the division in the health agency your utility works with the most?

1. Communicable diseases
2. Environmental health
3. Epidemiology, or
4. Infectious disease
5. [Do not read] Other [Specify]
8. [Do not read] Don't know

Q11. Is there a specific person in your utility who is responsible for coordination with the health agency in your service area?

1. Yes
0. No [Skip to Q13]
8. Don't know [Skip to Q13]

Q12. Which best describes his/her job title?

1. Communications/Public Relations specialist
2. Customer service
3. Engineer or technician
4. General Manager (chief of entire organization)
5. Laboratory staff
6. Manager (In charge of one unit) or,
7. System operator
8. [Do not read] Don't know
9. [Do not read] Other [specify] ___________

Q13. Has your utility ever worked *with* a public health department or agency on drinking water communications?

1. Yes
0. No [Skip to Q22]
8. Don't know [Skip to Q22]

Q14. Does your utility have a formal agreement on drinking water communications, such as a memorandum of understanding, with a state or local health agency?

1. Yes
0. No
8. Don't know

Q15. In the past 5 years (from 1999 to the present), has your utility worked with a health agency on communications about drinking water with each other?

1. Yes
0. No
8. Don't know

Q16. Which of the following best describes the nature of your utility's most recent collaboration with a health agency? [read]

1. Consumer Confidence Reports
2. Education or training session
3. Meeting or conference
4. Visual communications, or
5. Written communications
6. [Do not read] Other [Specify]

Q17. Which of the following were the utility communication goals for the most recent collaboration? [read, check all that apply]

	Goal	Yes	No	Don't Know
1	Emergency Planning			
2	Information exchange between the utility and health department			
3	Information exchange with community, or			
4	Relationship building			
5	[Do not read] Other [Specify]			

Q18. Would you describe your utility's most recent collaboration with a health agency as: [read]

1. Very successful
2. Successful
3. Minimally successful, or
4. Not successful
8. [Do not read] Don't know

Q19. In the past 5 years has your utility worked with a health agency on public communications about drinking water in your service area?

1. Yes
0. No [Skip to Q22]
8. Don't know [Skip to Q22]

Q20. In the past five years, has your utility worked to communicate with susceptible populations, for example the elderly, people with HIV or children?

1. Yes
0. No [Skip to Q22]
8. Don't know [Skip to Q22]

Q21. Did you work with a public health agency in communicating with susceptible populations?

1. Yes
0. No
8. Don't know

Q22. Do you work with public health agencies in calling boil water alerts?

1. Yes
0. No
8. Don't know

Q23. Please tell me which of the following methods your service area uses to communicate with the public about boil water alerts. [read, check all that apply]

	Method	Yes	No	Don't Know
1	Door Hangers			
2	E-mail			
3	Fax			
4	Newspapers			
5	Phone			
6	Radio			
7	TV, or			
8	[Do not read] Other [Specify]			

Q24. Are there special communications to susceptible populations during boil water alerts, or other emergency situations?

1. Yes
0. No
8. Don't know

Q25. Are boil water alerts communicated in languages other than English?

1. Yes
0. No
8. Don't know

Q26. Has your agency worked with utilities in developing emergency response plans as it relates to water security?

1. Yes
0. No
8. Don't know

Q27. Would you describe your utility's relationship with your local public health agency as: [read]

1. Excellent
2. Good
3. Fair, or
4. Poor

Q28. Which of the following do you feel would most improve your water utility's relationship with public health agencies? [read]

1. More frequent communication
2. Better quality of communications
3. More collaboration
4. Better interpersonal relationships
5. Better understanding of roles, or
6. Better understanding of needs and limitations
7. [Do not read] Other [Specify]
8. [Do not read] Don't know

Utility Communication With Health Care Providers

Now, I would like to ask you questions about communications with health care providers.
[Read if necessary: health care providers give direct health care, and include physicians, nurses, registered dieticians, and other allied health professions.]

Q29. Has your utility ever worked on drinking water issues with the health care providers in your service area?
1. Yes
0. No [Skip to Q46]
8. Don't know [Skip to Q46]

Q30. Which of the following best describes how often your utility communicates with health care providers about drinking water? [read]
1. Daily
2. Weekly
3. Monthly
4. Quarterly
5. Yearly or less
8. [Do not read] Don't know

Q31. Which types of health care providers? [read, check all that apply]

		Yes	No	Don't Know
1	Dentists			
2	Dieticians or Nutritionists			
5	Nurses, or			
6	Physicians			
8	[Do not read] Other [Specify]			

Q32. Does your utility communicate with medical specialists on drinking water issues?
1. Yes
0. No [Skip to Q34]
8. Don't know [Skip to Q34]

Q33. Does your utility communicate with: [read]?

		Yes	No	Don't Know
1	AIDS Specialists			
2	Dentists			
3	Emergency Department Staff			
4	Environmental/Occupational Medicine Specialists			
5	Gastroenterologists			
6	Geriatricians			
7	Infectious Disease Specialists			
8	Nephrologists/Kidney Specialists			
9	OB/GYNs, or			
10	Pediatricians			
11	[Do not read] Other [Specify]			

Q34. In the past five years, has your utility communicated with health care providers on: [read]

		Yes	No	Don't Know
1	Drinking water in general			
2	Drinking water emergencies (e.g. boil water alerts)			

Q35. In the past five years, has your water utility communicated with health care providers on about chemicals in drinking water?

1. Yes
0. No [Skip to Q37]
8. Don't know [Skip to Q37]

Q36. Has your utility communicated with health care providers about any of the following:

		Yes	No	Don't Know
1	Arsenic			
2	Chlorine/Disinfection Byproducts			
3	Fluoride			
4	Lead			
5	Nitrate			
6	Pesticides			
7	[Do not read] Other [Specify]			

Q37. In the past five years, has your utility communicated about microbial contaminants with health care providers?
1. Yes
0. No [Skip to Q39]
8. Don't know [Skip to Q39]

Q38. Did the utility communicate about any of the following:

		Yes	No	Don't Know
1	Coliform bacteria			
2	*Cryptosporidium*			
3	*Giardia*			
4	Viruses			
5	[Do not read] Other [Specify]			

Q39. Did your utility work with a public health agency in communicating with health care providers about contaminants?
1. Yes
0. No
8. Don't know

Q40. In the last five years, has your water utility worked with health care providers on communications for a susceptible population?
1. Yes
0. No [Skip to Q43]
8. Don't know [Skip to Q43]

Q41. Did the utility work with a public health agency in this communication?
1. Yes
0. No
8. Don't know

Q42. Which susceptible populations were targeted? [read, check all that apply]

		Yes	No	Don't Know
1	Cancer/Chemotherapy patients			
2	Children			
3	Chronic GI illness patients [Read if necessary: Chronic GI illness refers to conditions such as Crohn's disease or ulcerative colitis]			
4	Elderly			
5	Dialysis patients			
6	HIV/AIDS			
7	Immune Compromised, or			
8	Pregnant women			
9	[Do not read] Other [Specify]			

Q43. Thinking about your agency's most recent collaboration with health care providers, which of the following best describes the communication? [read]

1. Consultation
2. Consumer Confidence Reports
3. Education or training
4. Visual communication, or
5. Written communication
6. [Do not read] Other [Specify]

Q44. Which of the following were the utility communication goals for the most recent effort? [read, check all that apply]

		Yes	No	Don't Know
1	Community relations			
2	Emergency planning			
3	Information exchange, or			
4	Relationship building			
5	[Do not read] Other [Specify]			

Q45. Would you describe your utility's most recent collaboration with health care providers as: [read]

1. Very successful
2. Successful
3. Minimally successful, or
4. Not successful
8. [Do not read] Don't know

Q46. Did the water utility work with health care providers in developing your water security plan?

1. Yes
0. No
8. Don't know

Q47. Are health care providers included in your emergency response plans as it relates to water security?

1. Yes
0. No
8. Don't know

Q48. In an emergency, how does the utility communicate with health care providers?

		Yes	No	Don't Know
1	E-mail			
2	Fax			
3	Phone			
4	Through the Health Agency, or			
5	[Do not read] Other [Specify]			

Q49. Are there special communications used with health care providers for susceptible populations during boil water alerts?
1. Yes
0. No
8. Don't know

Q50. Overall, would you describe your utility's relationship with local health care providers as: [read]
1. Excellent
2. Good
3. Fair, or
4. Poor
8. [Do not read] Don't know

Q51. In your opinion, which method would be the best approach to communicate with health care providers? [read]
1. Conference presentations
2. Grand Rounds
3. Hospital staff meetings
4. Meetings in health care practices
5. Information sheets, or
6. Medical society presentations
7. [Do not read]Other[Specify]
8. [Do not read] Don't know

Q52. Which of the following do you feel would most improve your utility's relationship with health care providers? [read]
1. More frequent communication
2. Better quality communication, or
3. More collaboration
4. [Do not read] Other[Specify]
8. [Do not read] Don't know

Utility Demographics

I would like to ask a few questions about your service area.

Q53. How many people, not billing units, are served by your utility?
1. 1-500
2. 501-1000
3. 1001-10,000
4. 10,001-100,000
5. 100,001-500,000
6. 500,001-1,000,000
7. More than 1,000,000
8. [Do not read] Don't know

Q54. Would you categorize the utility service area as: [read]
1. Urban
2. Suburban
3. Rural
4. Mixed

Q55. How many political jurisdictions (i.e. city, municipality, county) are in the utility's service area?

1. 1
2. 2-5
3. 6-10
4. 11-20
5. 21 or more

Q56. How many full-time equivalent staff work in your utility?

1. 1-5
2. 6-10
3. 11-20
4. 21-50
5. 51-100
6. 101 or more

Personal Demographics

Finally, I would like to ask some broad questions about yourself.

Q57. Which of the following best describes your position within the water utility? [read]

1. Communications/Public Relations specialist
2. Customer service
3. Engineer or technician
4. General Manager
5. Laboratory staff
6. Manager (In charge of one unit), or
7. System operator
8. [Do not read] Other [Specify]

Q58. Which best describes your unit in the utility?

1. Administration
2. Communications
3. Customer relations
4. Engineering
5. General Manager's Office
6. Human Resources
7. Laboratory
8. Water quality or production
9. [Do not read] Other [Specify]

Q59. How many years have you worked in the water utility industry?

1. 1-5
2. 6-10
3. 11-20
4. 21 or more

Q60. What is the highest level of education you have completed?
1. High school graduate (Grade 12 or GED certificate)
2. Business, technical or vocational school AFTER high school
3. Some college, no 4 year degree
4. College graduate (B.S., B.A. or other 4 year degree)
5. Graduate level training or professional school after college (1-2 years, no degree)
6. Master's degree (ME, MS, MBA, etc)
7. Doctoral degree (PhD, MD)
8. Other advanced professional degree (JD, LLP)

Q61. Do you live in your utility's service area?
1. Yes
0. No

Q62. In which state is your utility located? [Do not read, check appropriate code]

AL	AK	AS	AZ	AR	CA	CO	CT	DC	DE	FL	GA
HI	IA	IL	IN	ID	KS	KY	LA	ME	MA	MD	MI
MN	MS	MO	MT	NE	NV	NH	NJ	NM	NY	NC	ND
OH	OK	OR	PA	RI	SC	SD	TN	TX	UT	VT	VA
WA	WV	WI	WY								

Those are all of our questions. Do you have any questions?

[Yes, answer as needed go to close.]

[No go to close]

[Close] Thank you very much for your time and assistance.

SURVEY QUESTIONS FOR HEALTH AGENCY PERSONNEL

Introduction:

Good morning/afternoon. I am (name) … from Opinion Search, calling on behalf of The George Washington University. I am calling in follow up to our recent letter about our survey on communication practices of local public health agencies (LPHA's).

Is this a good time to go through the questions?

[NO]
May I schedule a time to call you again?

Appointment (Day, Date, Time) ________________________________

[Yes]
Let's continue.

All the information you provide us will be kept strictly confidential. Only the project staff will have access to your responses with no personal information attached. No one else will have access to this information without your knowledge. The Institutional Review Board of The George Washington University Medical Center has reviewed and approved this questionnaire and the study procedures.

May I continue with the questions?

[If the participant wants to talk with the study's participant representative, s/he may call the GWUMC IRB at 202-994-2764. In that case, ask if you may call the participant back tomorrow or the next day.]

Q1. Gender [DO NOT ASK. Check off the answer.]
1. Male
2. Female

Thank you for agreeing to participate.

First, I would like to ask you about how your local public health agency works within government.

Governmental Structure and Organization

Q2. Which of the following best describes your jurisdiction?
1. City
2. County
3. State, or
4. [Do not read] Other

Q3. Does your agency have a role in regulating drinking water?
1. Yes, the local public health agency is the regulator
2. Yes, but the state is the primary regulator
0. No [Skip to Q5]
8. [Do not read] Don't know [Skip to Q5]

Q4. Which division is responsible for oversight?
1. Communicable Diseases
2. Community Health, or
3. Environmental Health
4. [Do not read] Other [Specify]
8. [Do not read] Don't know

Q5. Which agency or agencies regulate drinking water in your state?
[read, check all that apply]

		Yes	No	Don't Know
1	The Health Department or Agency			
2	The Department of Environmental Quality (DEQ) or Natural Resources (DNR)			
3	The Energy Department, or			
4	A Public Utility Commission			
5	[Do not read] Other [Specify]			

Public Health Communications With Water Utilities

Now I would like to ask you about how the agency and water utility communicate in your jurisdiction.

[Read if necessary: Communication refers to both formal and informal contacts where information is exchanged]

Q7. Which of the following best describes how often your agency communicates with water utilities? [read]
1. Daily
2. Weekly
3. Monthly
4. Quarterly
5. Yearly, or less
6. Never [Skip to Q11]
7. [Do not read] Other

Q8. When your agency communicates with water utilities, which of the following methods are used: [read, check all that apply]

		Yes	No	Don't Know
1	Conferences			
2	E-mail			
3	Fax			
4	Mail			
5	Meetings, or			
6	Phone			
7	[Do not read] Other [specify]			

Q9. Is there a specific person in your agency who is responsible for coordination with the water utilities in your jurisdiction?
1. Yes
0. No [Skip to Q11]
8. Don't know [Skip to Q11]

Q10. Which best describes his/her job title? [read]
1. Director of the agency (Manages all units)
2. State Epidemiologist
3. Staff level epidemiologist
4. Health officer (MD)
5. Program manager (Manages one or a few units), or
6. Nurse
7. [Do not read] Other [Specify]
8. [Do not read] Don't know

Q11. Has your agency ever worked with a water utility on drinking water communications?
1. Yes
0. No [Skip to Q20]
8. Don't know [Skip to Q20]

Q12. Does your agency have a formal agreement on drinking water communications, such as a memorandum of understanding, with a water utility?
1. Yes
0. No
8. Don't know

Q13. In the past 5 years (from 1999 to the present), has your agency worked with a water utility on communications about drinking water WITH EACH OTHER in your service area?
1. Yes
0. No [Skip to Q18]
8. Don't know [Skip to Q18]

Q14. Which of the following best describes the nature of your agency's most recent collaboration with a water utility? [read]

1. Consumer Confidence Reports
2. Education or training
3. Meeting or conference
4. Visual communications, or
5. Written communications
6. [Do not read] Other [specify]

Q15. Which of the following were the agency communication goals for the most recent collaboration? [read, check all that apply]

	Goal	Yes	No	Don't Know
1	Emergency Planning			
2	Information exchange between the utility and health department			
3	Information exchange with community, or			
4	Relationship building			
5	[Do not read] Other [specify]			

Q16. Would you describe your agency's most recent work with a water utility as: [read]

1. Very successful
2. Successful
3. Minimally successful, or
4. Not successful
8. [Do not read] Don't know

Q17. In the past 5 years has your agency worked with a water utility on public communications about drinking water in your jurisdiction?

1. Yes
0. No
8. Don't know

Q18. In the past five years, has your agency worked to communicate with susceptible populations, for example the elderly, people with HIV or children?

1. Yes
0. No [Skip to Q20]
8. Don't know [Skip to Q20]

Q19. Did your agency work with a water utility in communicating with susceptible populations?

1. Yes
0. No
8. Don't know

Q20. Does your agency work with water utilities in calling boil water alerts?

1. Yes
0. No
8. Don't know

Q21. Please tell me which of the following methods your jurisdiction uses to communicate with the public about boil water alerts. [read, check all that apply]

1	Method	Yes	No	Don't Know
2	Door Hangers			
3	E-mail			
4	Fax			
5	Newspapers			
6	Phone			
7	Radio, or			
8	TV			
9	[Do not read] Other [Specify]			

Q22. Are there special communications to susceptible populations during boil water alerts, or other emergency situations?

1. Yes
0. No
8. Don't know

Q23. Are boil water alerts communicated in languages other than English?

1. Yes
0. No
8. Don't know

Q24. Has your agency worked with utilities in developing emergency response plans as it relates to water security?

1. Yes
0. No
8. Don't know

Q25. Would you describe your agency's relationship with your local water utilities as: [read]

1. Excellent
2. Good
3. Fair, or
4. Poor

Q26. Which of the following do you feel would most improve your agency's relationship with water utilities?

1. More frequent communication
2. Better quality of communications
3. More collaboration
4. Better interpersonal relationships
5. Better understanding of roles, or
6. Better understanding of needs and limitations
7. [Do not read] Other [Specify]
8. [Do not read] Don't know

Public Health Communications With Health Care Providers

Now I would like to ask you about communications with health care providers.

[Read if necessary: health care providers give direct health care, and include physicians, nurses, registered dieticians, and other allied health professions.]

Q27. Has your agency ever worked on drinking water issues with the health care providers in your jurisdiction?

1. Yes
0. No [Skip to Q44]
8. Don't know [Skip to Q44]

Q28. Which of the following best describes how often your agency communicates with health care providers about drinking water?

1. Daily
2. Weekly
3. Monthly
4. Quarterly
5. Yearly, or less
6. [Do not read] Other [Specify]

Q29. Which types of health care providers? [read, check all that apply]

		Yes	No	Don't Know
1	Dentists			
2	Dieticians/Nutritionists			
3	Nurses, or			
4	Physicians			
5	[Do not read] Other [specify]			

Q30. Does your agency communicate with medical specialists on drinking water issues?

1. Yes
0. No [Skip to Q32]
8. Don't know [Skip to Q32]

Q31. Does your utility communicate with: [read]

		Yes	No	Don't Know
1	AIDS Specialists			
2	Dentists			
3	Emergency Department Staff			
4	Environmental/Occupational Medicine Specialists			
5	Gastroenterologists			
6	Geriatricians			
7	Infectious Disease Specialists			
8	Nephrologists/Kidney Specialists			
9	OB/GYNs, or			
10	Pediatricians			
11	[Do not read] Other [specify]			

Q32. In the past five years, has your agency communicated with health care providers on: [read check all that apply]

		Yes	No	Don't Know
1	Drinking water in general			
2	Drinking water emergencies (e.g. boil water alerts)			

Q33. In the past five years, has your agency communicated with health care providers about chemicals in drinking water?

1. Yes
0. No [Skip to Q35]
8. Don't know [Skip to Q35]

Q34. Has the agency communicated with health care providers about any of the following: [read check all that apply]

		Yes	No	Don't Know
1	Arsenic			
2	Chloramine Disinfection			
3	Disinfection Byproducts/Chlorine			
4	Fluoride			
5	Lead			
6	Nitrate, or			
7	Pesticides			
9	[Do not read] Other [specify]			

Q35. In the past five years, has your agency communicated about microbial contaminants with health care providers?
1. Yes
0. No [Skip to Q37]
8. Don't know [Skip to Q37]

Q36. Did the agency communicate about any of the following:[read check all that apply]

		Yes	No	Don't Know
1	Coliform bacteria			
2	*Cryptosporidium*			
3	*Giardia*, or			
4	Viruses			
5	[Do not read] Other [specify]			

Q37. Did your agency work with a drinking water utility in communicating with health care providers about drinking water contaminants?
1. Yes
0. No
8. Don't know

Q38. In the last five years, has your agency worked with health care providers on communications for a susceptible population?
1. Yes
0. No [Skip to Q41]
8. Don't know [Skip to Q41]

Q39. Did your agency work with a drinking water utility in this communication?
1. Yes
0. No
8. Don't know

Q40. Which susceptible populations were targeted? [read, check all that apply]

		Yes	No	Don't Know
1	Cancer or Chemotherapy patients			
2	Children			
3	Chronic GI illness patients [Read if necessary: Chronic GI illness refers to conditions such as Crohn's disease or ulcerative colitis]			
4	Elderly			
5	Dialysis patients			
6	HIV/AIDS			
7	Immune Compromised, or			
8	Pregnant women			
9	[Do not read] Other [specify]			

Q41. Thinking about your agency's most recent collaboration with health care providers, which of the following best describes the communication? [read]
1. Consultation
2. Consumer Confidence Reports
3. Education or training
4. Visual communication, or
5. Written communication
6. [Do not read] Other [specify]

Q42. Which of the following were the agency communication goals for the most recent effort? [read, check all that apply]

		Yes	No	Don't Know
1	Community relations			
2	Emergency planning			
3	Information exchange, or			
4	Relationship building			
5	[Do not read] Other [specify]			

Q43. Would you describe your agency's most recent collaboration with health care providers as: [read]
1. Very successful
2. Successful
3. Minimally successful, or
4. Not successful
5. [Do not read] Don't know

Q44. Did the agency work with health care providers in developing your emergency response plans as it relates to water security?
1. Yes
0. No
8. Don't know

Q45. Are health care providers included in your emergency response plans focused on water security?
1. Yes
0. No
8. Don't know

Q46. In an emergency, how does your agency communicate with health care providers? [read]

		Yes	No	Don't Know
1	E-mail			
2	Fax, or			
3	Phone			
4	[Do not read] Other [specify]			

Q47. Are there special communications used with health care providers for susceptible populations during boil water alerts?

1. Yes
0. No
8. Don't know

Q48. Overall, would you describe your agency's relationship with local health care providers as: [read]

1. Excellent
2. Good
3. Fair, or
4. Poor
8. [Do not read] Don't know

Q49. In your opinion, which method would be the best approach to communicate with health care providers? [read]

1. Conference presentations
2. Grand Rounds
3. Hospital staff meetings
4. Meetings **in** health care practices [in-services]
5. Information sheets [fact sheets]
6. Medical society presentations
7. [Do not read] Other [Specify]

Q50. Which of the following do you feel would most improve your agency's relationship with health care providers? [read]

1. More frequent communication
2. Better quality communication, or
3. More collaboration
4. [Do not read] Other [Specify]
8. [Do not read] Don't know

Health Agency Demographics

Now I would like to ask a few questions about your jurisdiction.

Q51. How many people live in your jurisdiction?

1. 1-500
2. 501-1000
3. 1001-10,000
4. 10,001-100,000
5. 100,001-500,000
6. 500,001-1,000,000, or
7. More than 1,000,000
8. [Do not read] Don't know

Q52. Would you describe your jurisdiction as [read]

1. Urban
2. Suburban
3. Rural, or
4. Mixed

Q53. How many water utilities are in the agency's jurisdiction?
1. 1
2. 2-5
3. 6-10
4. 11-20
5. 21 or more
8. [Do not read] Don't know

Q54. How many full-time equivalent staff work in your agency?
1. 1-5
2. 6-10
3. 11-20
4. 21 or more
8. [Do not read] Don't know

Personal Demographics

Finally, I would like to ask you some broad questions about yourself.

Q55. Which best describes your position within the health department? [read]
1. Director (manages several units)
2. Epidemiologist
3. Health officials (MD, head of agency)
4. Nurse
5. Program manager (manages one unit), or
6. Sanitarian/Inspector
7. [Do not read] Other [Please specify]

Q56. Which best describes your unit within the agency?
1. Communicable Diseases
2. Community Health or,
3. Environmental Health
4. [Do not read] Other [Specify] __________

Q57. How many years have you worked in public health?
1. 1-5
2. 6-10
3. 11-20 or
4. 21 or more

Q58. What is the highest level of education you have completed?
1. High school graduate (Grade 12 or GED certificate)
2. Business, technical or vocational school AFTER high school
3. Some college, no 4 year degree
4. College graduate (B.S., B.A. or other 4 year degree)
5. Graduate level training or professional school after college (1-2 years, no degree)
6. Master's degree (ME, MS, MBA, etc)
7. Doctoral degree (PhD, MD), or
8. Other advanced professional degree (JD, LLP)

Q59. Do you live in your agency's jurisdiction?

1. Yes
0. No

Q60. In which state is your agency located? [Do not read, check code]

AL	AK	AS	AZ	AR	CA	CO	CT	DC	DE	FL	GA
HI	IA	IL	IN	ID	KS	KY	LA	ME	MA	MD	MI
MN	MS	MO	MT	NE	NV	NH	NJ	NM	NY	NC	ND
OH	OK	OR	PA	RI	SC	SD	TN	TX	UT	VT	VA
WA	WV	WI	WY								

Those are all of our questions. Do you have any questions?

[Yes, answer as needed go to close.]

[No go to close]

[Close] Thank you very much for your time and assistance.

HEALTH CARE PROVIDER INTERVIEW GUIDE

Introduction:

Good morning/afternoon. I am (name) … calling from The George Washington University. I am calling in follow up to our recent letter about our survey on communication practices of clinicians in regard to drinking water.

Is this a good time to go through the questions? This should take about 15 minutes.

[NO]
May I schedule a time to call you again?

Appointment (Day, Date, Time) ________________________________

[Yes]
Let's continue.

All the information you provide us will be kept strictly confidential. Only the project staff will have access to your responses with no personal information attached. No one else will have access to this information without your knowledge. The Institutional Review Board of The George Washington University Medical Center has reviewed and approved this questionnaire and the study procedures.

May I continue with the questions?

[If the participant wants to talk with the study's participant representative, s/he may call the GWUMC IRB at 202-994-2764. In that case, ask if you may call the participant back tomorrow or the next day.]

Q1. Gender [DO NOT ASK. Check off the answer.]
1. Male
2. Female

The first questions are about general issues on waterborne disease in clinical practice.

Etiology and Disease Knowledge

Q2. When taking a patient history, have you ever asked about potential waterborne disease exposure?
1. Yes
0. No [Skip to Q4]
8. Don't know [Skip to Q4]

Q3. What prompts you to ask waterborne disease-related questions in a patient history?

		Yes	No	Don't Know
1	Emergency situations (boil water alerts, weather)			
2	Laboratory Results			
3	Patient concerns			
4	Symptoms			
5	[Do not read] Other [Specify]			

Q4. Have you **ever** diagnosed a disease or condition that you felt was caused by drinking water exposure?
1. Yes
0. No
8. Don't know

Q5. In the past five years, have you looked for information about waterborne disease?
1. Yes
0. No [Skip to Q7]
8. Don't know [Skip to Q7]

Q6. Which of the following sources of information have you used? [read]

	Source	Yes	No	Don't Know
1	CDC			
2	Colleagues			
3	EPA			
4	Local Health Department			
5	Medline or other on-line searches			
6	Medical Journals			
7	State health department			
8	Water utilities			
9	[Do not read] Other [Specify]			

Public Health Communications

Next, I would like to discuss how you work with the local public health agency.

Q7. When you communicate with health agencies, which of the following methods do you use: [read, check all that apply]

		Yes	No	Don't Know
1	Conferences			
2	E-mail			
3	Fax			
4	Mail [snail]			
5	Meetings			
6	Phone			
7	[Do not read] Other [specify]			

Q8. Who do you rely on to contact the health department about reportable conditions?

		Yes	No	Don't Know
1	Contractor			
2	Laboratory			
3	Self			
4	Staff			
5	[Do not read] Other [Specify]			

Q9. Have you ever contacted a health agency about drinking water-related health issues?
1. Yes
0. No [Skip to Q15]
8. Don't know [Skip to Q15]

Q10. How frequently do you communicate with the health department about drinking water concerns?
1. Daily – Weekly
2. Monthly
3. Quarterly
4. Yearly or less
5. Never
8. [Do not read] Other

Q11. Is there a specific department in the health agency you work with on drinking water issues?
1. Yes
0. No [Skip to Q15]
8. Don't know [Skip to Q15]

Q12. Which best describes the unit within the agency? [read]
1. Communicable Diseases
2. Community Health
3. Environmental Health
4. Laboratory
5. [Do not read] Other [Specify]
8. [Do not read] Don't know

Q13. Is there a specific person you work with in the health department on drinking water issues?
1. Yes
0. No [Skip to Q15]
8. Don't know [Skip to Q15]

Q14. Which best describes the position or the person within the health department that you contact most often?
1. Director (manages several units)
2. Epidemiologist
3. Health official (MD, or head of agency)
4. Nurse
5. Program manager (manages one unit)
6. Sanitarian/Inspector
8. [Do not read] Don't know

Q15. In the past five years, has your local public health agency or department ever contacted you to collaborate on drinking water issues?
1. Yes
0. No [Skip to Q18]
8. Don't know [Skip to Q18]

Q16. Were you able to work with the health department?
1. Yes
0. No
8. Don't know

Q17. Was a water utility involved in the collaboration?
1. Yes
0. No
8. Don't know

Q18. In the past five years, have you **received** communications about drinking water-related health issues from your local or state public health agencies?
1. Yes
0. No [Skip to Q25]
8. Don't know [Skip to Q25]

Q19. Was this information from health agencies at the state or local level?
1. Local
2. State
3. Other
8. Don't know

Q20. Which of the following activities have health agencies used to communicate about drinking water and health with you? [read]

		Yes	No	Don't Know
1	Conference Presentations			
2	Direct Communication			
3	Grand Rounds			
4	Hospital Staff Meetings			
5	Meetings **in** health care practices			
6	Information sheets			
7	Medical Society Presentations			
8	[Do not read] Other [Specify]			

Q21. Would you describe your MOST recent collaboration with public health agencies as: [read]
1. Very successful
2. Successful
3. Minimally successful
4. Not successful
8. [Do not read] Don't know

Q22. Do you feel your most recent collaboration was effective in addressing drinking water and health issues for a clinical setting?
1. Yes
0. No
8. Don't know

Q23. What do you feel is **the most successful** type of collaboration between clinicians and health agencies? [read]
1. Community education
2. Patient education
3. Training or presentations between the health agency and health care providers
4. Emergency situations
5. [Do not read] Other [Specify]

Q24. Do you feel the **least successful** type of collaboration between clinicians and health agencies is? [read]
1. Community education
2. Patient education
3. Training or presentations between the health agency and health care providers
4. Emergency situations
8. [Do not read] Other [Specify]

Q25. Have you worked with your local public health agencies on emergency response plans that included drinking water issues?
1. Yes
0. No
8. Don't know

Q26. In the past five years, have you received any referrals or questions from the local public health agency regarding drinking water and health?
1. Yes
0. No
8. Don't know

Q27. Would you describe your relationship with your local public health agency as: [read]
1. Excellent
2. Good
3. Fair
4. Poor

Q28. Which of the following do you feel would MOST improve your relationship with health agencies? [read]
1. More frequent communication
2. Better quality communication
3. More collaboration
4. [Do not read] Other [specify] __________
8. [Do not read] Don't know

Q29. Would you be willing to collaborate on communications with public health agencies on drinking water topics?
1. Yes
0. No [Skip to Q31]
8. Don't know [Skip to Q31]

Q30. How often do you feel you could work with a public health agency on drinking water-related health issues?
1. Weekly
2. Monthly
3. Quarterly
4. Yearly, or less
5. Never
6. [Do not read] Other [Specify]

Q31. Which of the following agency or agencies regulate drinking water in your state?
[read, check all that apply]

		Yes	No	Don't Know
1	Health Department or Agency			
2	Department of Environmental Quality (DEQ) or Natural Resources (DNR)			
3	Energy Department			
4	Public Utility Commission			
5	[Do not read] Other [Specify]			

Q37. Do you feel there are other opportunities for collaborating with the health department?

Q38. What if any barriers do you experience in collaborating with the health department?

Utility Communications

Now, I would like to ask you about collaborations or work you may have done with drinking water utilities.

Q39. Has a water utility ever contacted you to collaborate on drinking water issues?
1. Yes
0. No [Skip to Q43]
8. Don't know [Skip to Q43]

Q40. Were you able to work with the water utility?
1. Yes
0. No [Skip to Q43]
8. Don't know [Skip to Q43]

Q41. Which of the following best describes the type of collaboration?

		Yes	No	Don't Know
1	Conferences			
2	Emergency Planning			
3	Grand Rounds			
4	Information sheets			
5	Hospital staff meetings			
6	Meetings **in** health care practices			
7	Medical society presentations			
8	[Do not read] Other [specify]			

Q42. Was a public health agency part of this collaboration?
1. Yes
0. No
8. Don't know

Q43. In the past five years, have you received any communications from a water utility, such as Consumer Confidence Reports, special notices, etc?
1. Yes
0. No [Skip to Q46]
8. Don't know [Skip to Q46]

Q44. Which methods did the water utility use for these communications?

		Yes	No	Don't Know
1	Conferences			
2	E-mail			
3	Fax			
4	Mail			
5	Meetings			
6	Phone			
7	[Do not read] Other [specify]			

Q45. Did you feel that these communications were effective in addressing drinking water and health issues for a clinical setting?
1. Yes
0. No
8. Don't know

Q46. Have you collaborated with water utilities on drinking water emergency plans, such as boil water alerts or water security plans?
1. Yes
0. No
8. Don't know

Q47. In the last five years, have you contacted a water utility about drinking water health concerns?
1. Yes
0. No
8. Don't know

Q48. Would you be willing to work with a water utility on drinking water and health issues?
1. Yes
0. No [Skip to Q50]
8. Don't know [Skip to Q50]

Q49. How frequently do you feel you could work with a drinking water utility?
1. Weekly
2. Monthly
3. Quarterly
4. Yearly, or less
5. Never
6. [Do not read] Other [Specify]

Q50. Are there any other opportunities for collaborating with water utilities you would like to discuss?

Q51. What, if any, barriers do you experience in collaborating with water utilities?

Patient Communications

Now I would like to move on to ask general questions about your practice and drinking water issues.

Q52. How frequently do your patients ask you about drinking water related topics?
1. Daily
2. Weekly
3. Monthly
4. Quarterly
5. Yearly, or less
6. Never [Skip to Q56]
7. [Do not read] Other [Specify]

Q53. Do patients ask about drinking water-related diseases or contaminants, such as

		Yes	No	Don't Know
1	Arsenic			
2	Chemicals [general]			
3	Chlorine			
4	Diarrhea			
5	*Giardia*			
6	Fluoride			
7	Microbes [general]			
8	Pesticides			
9	[Do not read] Other [Specify]			

Q54. Have patients **ever** asked you about information from their Consumer Confidence Reports?
1. Yes
0. No
8. Don't know

Q55. In your experience, have any of the following communications prompted your patients to ask drinking water related questions?

	Method	Yes	No	Don't Know
1	Boil Water Alerts			
2	Newspaper articles			
3	Radio stories			
4	TV stories			
5	[Do not read] Other [specify]			

Now I would like to ask you about your medical training and professional organizations.

Q56. Have you **ever** received training in environmental medicine?
1. Yes
0. No [Skip to Q58]
8. Don't know [Skip to Q58]

Q57. Do you feel that you received enough training on drinking water related health issues to adequately address your patients' concerns?
1. Yes
0. No
8. Don't know

Q58. Are you a member any of the following medical organizations or associations?

	Org/Assoc	Yes	No	Don't Know
1	American College of Preventive Medicine			
2	American College of Occupational and Environmental Medicine			
3	American Medical Association			
4	Association of Occupational and Environmental Clinics			
	Local Medical Society			
5	Physicians for Social Responsibility			
	State Medical Society			
6	[Do not read] Other [Specify]			

Q59. Have you used any drinking water-related communications from any of these organizations or associations?

1. Yes
0. No
8. Don't know

Next I would like to ask some questions about specific drinking water topics.

Q60. Has a public health agency communicated to you about: [read]

		Yes	No	Don't Know
1	Drinking water in general			
2	Drinking water emergencies (e.g. boil water alerts)			

Q61. Has a drinking water utility communicated to you about: [read]

		Yes	No	Don't Know
1	Drinking water in general			
2	Drinking water emergencies (e.g. boil water alerts)			

Q62. Which of the following chemicals found in drinking water are of interest to you for your clinical practice: [read]

		Yes	No	Don't Know
1	Arsenic			
2	Chloramine Disinfection			
3	Disinfection Byproducts/Chlorine			
4	Fluoride			
5	Lead			
6	Nitrate			
7	Pesticides			
8	None			
9	[Do not read] Other [specify]			

Q63. Which of the following microbial contaminants are of interest for you in your clinical work?

		Yes	No	Don't Know
1	Coliform bacteria			
2	*Cryptosporidium*			
3	*Giardia*			
4	Viruses			
5	None			
6	[Do not read] Other [specify]			

Q64. Are any of the following susceptible populations a concern to you in regard to drinking water and health? [read]

		Yes	No	Don't Know
1	Cancer or Chemotherapy patients			
2	Children			
3	Chronic GI illness patients			
4	Elderly			
5	Dialysis patients			
6	HIV/AIDS			
7	Immune Compromised			
8	Parents, or			
9	Pregnant women			
10	[Do not read] Other [specify]			

Demographics

Now I am going to ask a few questions about the area where you practice.

Q65. What is the population in the locality where you practice?
1. 1-500
2. 501-1000
3. 1001-10,000
4. 10,001-100,000
5. 100,001-500,000
6. 500,001-1,000,000
7. More than 1,000,000
8. Don't know

Q66. Would you describe your practice area as: [read]
1. Urban
2. Suburban
3. Rural
4. Mixed

Q67. How many water utilities are in the locality where you practice?
1. 1
2. 2-5
3. 6-10
4. 11-20
5. 21 or more
8. Don't know

Q68. Which best describes your clinical position in your practice?
1. Physician
2. Nurse
3. Other health care provider
4. Researcher
5. Administrator/Office Manager
6. [Do not read] Other [Specify]

Q69. How many clinicians (physicians, nurses, other direct health care providers) work in your practice?
1. 1-5
2. 6-10
3. 11-20
4. 21 or more
8. Don't know

Personal Demographics

Finally, I would like to ask some general questions about yourself and your practice.

Q70. Which best describes your type of medical practice?
1. Environmental and occupational medicine
2. Occupational medicine
3. Pediatrics
4. Preventive medicine
5. Primary care
6. Rehabilitative medicine
7. [Do not read] Other [Specify]

Q71. How many years have you worked in clinical settings?
1. 1-5
2. 6-10
3. 11-20
4. 21 or more

Q72. What is the highest level of education you have completed?
1. College graduate (B.S., B.S.N. or other 4 year degree)
2. Professional school after college (PA, MPT, NP)
3. Master's degree (MSN, MS, MBA, etc)
4. Medical Degree (MD or DO)
5. Doctoral degree (PhD, DrPH)
6. Registered clinical professional (RN, RD)
7. [Do not read] Other [Specify]

Q73. Do you live in your practice area?
1. Yes
0. No

Q74. In which state do you practice?

AL	AK	AS	AZ	AR	CA	CO	CT	DC	DE	FL	GA
HI	IA	IL	IN	ID	KS	KY	LA	ME	MA	MD	MI
MN	MS	MO	MT	NE	NV	NH	NJ	NM	NY	NC	ND
OH	OK	OR	PA	RI	SC	SD	TN	TX	UT	VT	VA
WA	WV	WI	WY								

Those are all of our questions. Do you have any questions?

[Yes, answer as needed go to close.]

[No go to close]

[Close] Thank you very much for your time and assistance.

APPENDIX C
SURVEY AND INTERVIEW DATA

Table C.1
Water utility demographics

Participant demographics	
Male	87%
Working in the industry more than 10 years	84%
Manager	72%
Lives in the utility service area	65%
Utility demographics	
The participant works at a utility that:	
Is publicly owned	92%
Has 1-5 political jurisdictions in the utility jurisdiction	73%
Is governed by elected officials	70%
Serves up to 100,000 customers	67%
Has more than 20 employees	53%
Is in a mixed jurisdiction (urban/suburban/rural)	42%
Is governed by appointed officials	27%
Has a community advisory board	15%
The participant works at a utility that works with:	
Local PHAs	89%
State PHAs	88%
Federal agencies	64%
Other	
In their area, drinking water is regulated by	
State environmental agencies	76%
S/LPHA	61%

Table C.2
Local public health agency demographics

Participant Demographics	
Male	74%
Manager	71%
Working in the profession for more than 20 years	48%
Director of two or more organizational units	46%
Agency Demographics	
The participant works at an agency that:	
Is a county-level organization	67%
Has a role in regulating water utilities	60%
Has more than 5 water utilities in jurisdiction	59%
Serves up to 100,000 customers	56%
Has more than 20 employees	53%
The participant's agency is located in the:	
Midwest	42%
Northeast	21%
Southeast	20%
West	17%
Other	
The Environmental Health Unit is responsible for drinking water regulation	67%

Table C.3
Results of LPHA and utility surveys: communication between S/LPHAs and water utilities

	Utility response	LPHA response
Frequency of contact with the S/LPHA or utility		
Weekly or more	30%	32%
Monthly	46%	18%
Less than monthly	24%	46%
Mode of communication with the S/LPHA or utility		
Phone	96%	93%
Mail	87%	75%
E-mail	85%	67%
Faxes	74%	68%
Meetings	77%	69%
Organization has a designated official for communicating with the S/LPHA or utility	79%	69%
Organization has a formal agreement with the S/LPHA or utility for drinking water communications	33%	18%
Organization has worked with S/LPHA or utility on communications about drinking water in the last 5 years	85%	91%
On CCRs?	36%	26%
On other communications?	64%	74%
In past 5 years, LPHA and utility have worked together on:		
A boil water alert	85%	83%
Emergency response plans	89%	69%
Relationship with utility/LPHA is good or excellent	87%	82%
Necessary for improving relationship with S/LPHA or utility:		
Better understanding each other's needs and limitations	32%	21%
More frequent communication	18%	15%
More collaboration	17%	30%
A better understanding of each other's roles	15%	16%

Table C.4
Results of LPHA and utility surveys: communication with clinicians

	Utility response	LPHA response
Have ever worked with HCPs on drinking water issues	32%	54%
Communicate with HCPs once per year or less	74%	50%
People with whom communicated:		
Physicians	84%	95%
Nurses	39%	81%
Dentists	32%	52%
Among those that communicated, exchange was about:		
Lead	77%	73%
Chlorine	77%	50%
Coliform bacteria	85%	95%
Giardia	85%	77%
Cryptosporidium	90%	61%
Plan to communicate in emergency conditions via:		
Phone	79%	92%
Email	34%	84%
Relationships with HCPs are excellent or good	70%	83%

Table C.5
Related factors for interactions between water utilities and S/LPHAs

	Utilities	LPHAs
The larger the population (over 100,000) in the area served, the more likely they were to…		
Have ever worked with the other party on drinking water communications	$p<0.005$	$p=0.06$
Have worked with the other party on water security-related ERPs	$p<0.05$	$p=0.06$
Communicate with the other party weekly or more often	$p<0.05$	$p<0.01$
Have worked with the other party to communicate with susceptible subpopulations in the past five years	$p=0.05$	n.s.*
Have worked with the other party in calling boil water alerts	n.s.*	$p<0.10$
The larger the population (over 100,000) in the area served, the <u>less</u> likely they were to have designated a specific staff person to coordinate with the other party	$p<0.05$	n.s.*
Utility managers† were <u>more</u> likely than other utility respondents to report that the organization...		
had ever worked with a state PHA	$p=0.08$	—
had recently‡ participated in joint communication activity to produce CCR	$p<0.05$	—
had recently collaborated with S/LPHA on drinking water communication	$p<0.01$	—
LPHA managers were <u>more</u> likely than other LPHA respondents to say that the agency…		
had ever worked with a utility on a drinking water communication	—	$p=0.06$
had designated a specific person to communicate with utilities	—	$p<0.05$

*Not significant
†These same relationships were generally true with employees who had worked in the profession a long time. This would be expected since the rank of manager is related to the length of service in the profession.
‡In the past five years

Table C.6
Clinician interview data

Demographics	
Specialist in environmental medicine or environmental health services	90%
Physician	87%
In practice for more than 10 years	83%
Member of the American College of Occupational and Environmental Medicine	73%
Male	63%
Practices in an environmental health clinic	60%
Works in a practice with fewer than six clinicians	57%
Works in a clinic that serves an area with a population of over 500,000	50%
Information about patients	
Said patients asked questions about drinking water issues. Questions occurred:	90%
Daily to monthly	56%
In response to newspaper articles	77%
In response to stories on television	60%
In response to Internet searches	57%
In response to radio stories	50%
Said patients has asked for information on:	
Chemicals generally	93%
Mercury	85%
Lead	78%
Arsenic	70%
Chlorine	63%
Pesticides	63%
Microbes	59%
Giardia	52%
Fluoride	52%
CCRs	13%
Had asked patients about potential sources of waterborne disease	93%
Had diagnosed a water-related illness	83%
Said patient symptoms had prompted them to look for water-related etiologies	63%

(continued)

Table C.6 (Continued)

Water-health information	
Use the following tools to find water-health information	
On-line tools such as Medline	83%
Centers for Disease Control and Prevention	77%
Agency for Toxic Substances and Disease Registry	67%
Medical journals	67%
Relevant SPHA	60%
Professional colleagues	57%
Professional medical societies	53%
LPHAs	53%
Water utilities	27%
Had received general drinking water information from	
S/LPHAs	47%
Water utilities	33%
Had received emergency drinking water information from	
S/LPHAs	37%
Water utilities	13%
Want more information about	
arsenic, nitrates, lead, mercury, pesticides, coliform bacteria, *E. coli*, *Cryptosporidium*, and/or *Giardia*	>80%
MTBE, fluoride, disinfection byproducts, chlorine, and/or chloramines	43–67%

Table C.7
Results of clinician interviews: collaboration with S/LPHAs and water utilities

Collaboration with water utilities	
Had been contacted by a water utility for collaboration on a drinking water issue	10%
Collaboration with S/LPHAs	
Characterize relationships with LPHAs as excellent or good	66%
Had received a written communication, mail, or CCR from a utility	57%
Wanted more frequent communication or more collaboration with LPHAs	50%
Had worked with a S/LPHA on emergency response planning	30%
Mode of communication with the S/LPHA	
Phone	97%
E-mail	73%
Meetings	60%
Mail	57%
Have contacted a S/LPHA about a drinking water-related issue	63%
Had contacted a S/LPHA at least once year	97%
Had established a practice of contacting a specific program	50%
Had been contacted by a LPHA for a referral or expertise	30%
Had received information from the S/LPHA about…	
General drinking water issues	46%
Emergency drinking water issues	37%
Had been contacted by a LPHA about collaborating on drinking water issues	23%
Partnership was clinically relevant	100%
Collaboration was successful or very successful	71%

APPENDIX D
KEY QUOTES FROM THE WORKSHOP FOCUS GROUPS

Table D.1
Key points made by utility representatives

On three-way communications	"Each agency has different information practices; the public just sees it as confusion. It becomes a circus, and it can lead to real communication failures." "TRUST makes a huge difference on communications across all issues. It also helps if you can 'talk off the record' to work out some problems."
On public health agencies	"Personal connections and communications with the State Health Department are important. We constantly meet, and I can pick up the phone any time…the city and county health departments not so much." "Regular reporting works well, you have to make it work. There is very little communication beyond that." "We have a difficult time getting local health agencies to read reports. You need to get their interest; they don't have the focus we have." "Local health departments should get doctors to go to the health department for responses [to patients' questions about water]."
On clinicians	"The question is if it is appropriate for utilities to communicate with the medical community. We see it as a health department role." "We would like to have a network of information so doctors could be more accurate in answering their patients' questions." "We would like a vehicle to educate doctors that doesn't drive them nuts." "There is a reason to NOT educate physicians…We are not responsible for the water once it gets to a private home. By utilities talking to doctors, it's too big of a rush. Drinking water is a public health success story."

Table D.2
Key points made by LPHA staff

On water utilities	"The utility is anxious about how the local health agency is going to communicate elevated levels [of contaminants] with the public because it could increase public distrust of the utility."
	"The utility has an interest in not thinking something is a health concern."
On clinicians	"Our [LPHA] Director has been there [working with physicians] for 10 years, and she has really cultivated relationships. The health department is really working now ... to do district outreach to doctors…"
	"Our health officer goes out and speaks to future clinicians."
	"We have no communication system with clinicians, and yes with hospitals and infectious disease groups, that's about it, but not clinicians."
	"Capacity is a major issue."
	"Water is just one of many, many, many issues we need to reach clinicians on."
	"You have to set yourself up as a credible source. How you do that, I don't know."
	"One person responded, "Go out and meet people. There is no substitute for face meetings."
	"There are people, including physicians, that don't trust the government. And that is hard to overcome."
On other Stakeholders	There are "huge health disparities, and there are cultural groups that clearly still do not trust the government."
	"Lots of people think that water is sterile…[It's a] vague concern that they don't understand."
	"People still buy bottled water. Even those who cannot afford it. A lot of trust issues. Ten years have past [since an emergency situation] and people still distrust."

Table D.3
Key points made by clinicians

On three-way interactions	"I think the idea of three-way communication...is very good. But it would be OK for me if my local health department is the conduit between the water utility and me...The health department should be keeping an eye on the water and have a way to contact me quickly and let me know what I need to know." "If it's a doctor you feel like you can get through it quickly and you will know what you are saying..." "My bias is to speak to another physician but if it was a nurse and he or she was responsive and able to speak in the appropriate language that would be fine with me...but you need someone who will not brush you off...who can be dependable either to help give you an answer or help you along the way to find the answer...." "But it [what's important] is getting to the person that you think will have the information. I know I can get to a higher level in the health department versus the water company where you will wind up with a low level person – that would really waste my time." "But with resources strapped in this world we live in, if I had one nickel to spend on health care, I wouldn't spend it on water because there are more important things that threaten my patients' health..."
On water utilities	"It's a perception that the water system is intact. If there was a problem with infection, it would probably be food contamination; probably not water supply." "Our mindset is that we are obligated to individuals and the water utilities are obligated to the community... [so] the natural flow would be from the water utilities to us." "We expect them [utilities] to establish relationships...since it is one of them and thousands of us." "The liaison can not be a PR [public relations] person. It has to be somebody of substance...Someone that can speak my language..." "Couch information that patients are vulnerable and that would be prudent for us to understand the water system. So that would get us involved, more communication going." "One way for water utilities to reach physicians in the community is to hire physicians and health educators to put together brochures about water...Get into the [clinicians'] offices, bring to us literature that has been reviewed by someone we trust...that will elevate everyone's consciousness."
On public health agencies	"It goes back to the culture of physicians – figuring it out for yourself, doing it yourself, and basically not wanting a boss. That is a lot of the barriers of how we were trained, regardless of how long we have been doing this. It's that same mindset." "No one ever mentioned the water to me in medical school. You don't have an environmental medicine rotation... It is not in our culture to think along those lines." "When we realize we are all seeing the same thing, we go to the medical director and she calls the health department. They come down and we all figure it out together."

Table D.4
Key points made by others

On three-way interactions	"Nobody ever wants to admit guilt...If nobody is hurt, then you don't fess up." "People don't want to divulge information that they could get in trouble for divulging."
On water utilities	Collaborating with utilities "is an important relationship. We look for opportunities to do things together." "We went through hard times together, which helped to build personal relationships." "It was a long fight to get the utility involved...now they love it and suggest trainings..."
On public health agencies	"We're [academicians] paid to provide technical support to local health departments..."
On the public	"The public perceives quality issues as it relates to catastrophes. Otherwise, it [water] is taken for granted." "People have different needs." "You need a 'teachable moment' ... and a short one-liner [to get people's attention]." "Identify a person to communicate with other populations... [Find] a trusted person in the community. Find them before the crisis."

APPENDIX E
COST-BENEFIT SCORECARD EXAMPLES

Last Updated by__Pat Smith__
On (date) ___July 12, 2005_____

TOOL C1.2.1
Select the options of interest

The collaborators need to describe the baseline or current state of the world as the status quo, the alternative in which no changes are made, as well as a set of potential options, which should range from ones that require minimal change to ones that entail extensive, new policies, programs, and partnerships. One option should describe the ideal option, regardless of whether it requires extensive change or not. In this step, it is important to consider a wide range of possible approaches for addressing the problem.

EXAMPLE

Option	Description
Baseline	Status quo
Option A	Leverage existing partnerships; could include developing informal links with community organizations, the medical society, or a university to facilitate collaboration among sectors.
Option B	Develop legal tools for facilitating collaboration among public health agencies, utilities, and clinicians.
Option C	Develop an ongoing water-health communication campaign in which a representative at the public health agency is the lead, but works in informal collaboration with clinicians and partners at the water utility.

Last Updated by Pat Smith
On (date) July 12, 2005

TOOL C1.2.2
Key characteristics of each option

	BASELINE: Current situation and resources	A: Form/expand partnerships	B: Use legal tools to formalize relationships	C: Establish PHA as communication lead
Purpose	Dialog and common understanding	Share resources to address common issues; build interdependent system to address issues	Institutionalize system to ensure tasks are done	Provide long-term coordination to match current and emerging needs; limit duplication & ensure tasks are done
Structure				
Roles/Structure	Loosely defined	Central body for communication; roles somewhat defined	Central body of decision makers; defined roles	Mutually agreed process used in decision making; formal roles, time, & evaluation
Links	Loose/flexible	Links advisory, informal	Formalized links	Formal, written in work assignments
Resources	No new	Leverages existing resources	Develops new resources	Develops new resources with coordinated budgets across entities
Process				
Decision making	Only on an incident-driven or emergency basis	Ad hoc processes	Structured group methods – in central and subgroups	Ideas and decisions shared
Leadership	Low key	Facilitative leaders	Specific leader, identified	Formal structure; trust level high
Communication	Informal	Clear and frequent	Formal, within central group	Highly developed
Time	Present	1 year	2-3 years	3-5 years

Last Updated by Pat Smith
On (date) July 12, 2005

TOOL C2.2
Develop outcome categories

	Outcome category 1	Outcome category 2	Outcome category 3
Personnel	Inadequate – significant resources needed above current levels	Marginal – limited resources needed above current levels	Adequate – no additional resources needed above current levels
Equipment	Inadequate – significant resources needed above current levels	Marginal – limited resources needed above current levels	Adequate – no additional resources needed above current levels
Facilities	Inadequate – significant resources needed above current levels	Marginal – limited resources needed above current levels	Adequate – no additional resources needed above current levels
Supplies	Inadequate – significant resources needed above current levels	Marginal – limited resources needed above current levels	Adequate – no additional resources needed above current levels
Sustainability	Low – significant changes to policies & procedures to ensure sustainability	Moderate changes to policies & procedures needed to ensure sustainability	High – no changes to policies & procedures needed to ensure sustainability
Management support	Low – little or no support from top management	Moderate support from top management	High – significant support exists from top management
Constituency support	Low – no support from constituents	Moderate support from constituents	High – significant support from constituents
Timeliness	Low – collaboration provides no response or slow response to needs	Moderate response to affected needs; similar to usual response time	High – rapid response to population needs
Relevance to consumer needs	Low – no relevance to consumer's info needs	Moderate relevance to consumer's info needs	High – most relevant to consumer's info needs
Quality of information	Low – quality of information poor or unknown	Satisfactory information quality	High – provides the best possible information

Last Updated by Pat Smith
On (date) July 12, 2005

TOOL C2.3
Forecast impacts

	BASELINE: Current situation and resources	A: Form/expand partnerships	B: Use legal tools to formalize relation	C: Establish PHA as lead
Utility				
Personnel	Adequate	*Inadequate*	Marginal	Marginal
Equipment	Adequate	*Inadequate*	Marginal	Marginal
Facilities	Adequate	*Inadequate*	Marginal	Marginal
Sustainability	*Low*	Moderate	High	Moderate
Mngmt support	High	Moderate	*Low*	Moderate
Constit support	*Low*	High	Moderate	Moderate
Timeliness	*Low*	High	Moderate	Moderate
Relevance	*Low*	High	Moderate	Moderate
Info quality	*Low*	High	Satisfactory	Satisfactory
Public Health Agency				
Personnel	Adequate	Marginal	Marginal	*Inadequate*
Equipment	Adequate	Marginal	Marginal	*Inadequate*
Facilities	Adequate	Marginal	Marginal	*Inadequate*
Sustainability	*Low*	Moderate	High	Moderate
Mngmt support	High	Moderate	Moderate	*Low*
Constit support	*Low*	Moderate	Moderate	High
Timeliness	*Low*	Moderate	Moderate	High
Relevance	*Low*	Moderate	Moderate	High
Info quality	*Low*	Satisfactory	Satisfactory	High
Clinical Entity				
Personnel	Adequate	Marginal	Marginal	Marginal
Equipment	Adequate	Marginal	Marginal	Marginal
Facilities	Adequate	Marginal	Marginal	Marginal
Sustainability	*Low*	Moderate	High	Moderate
Mngmt support	High	Moderate	*Low*	Moderate
Constit support	*Low*	Moderate	Moderate	High
Timeliness	*Low*	Moderate	Moderate	High
Relevance	*Low*	Moderate	Moderate	High
Info quality	*Low*	Satisfactory	Satisfactory	High
Community Organizations				
Personnel	Adequate	Marginal	Marginal	Marginal
Equipment	Adequate	Marginal	Marginal	Marginal
Facilities	Adequate	Marginal	Marginal	Marginal
Sustainability	*Low*	Moderate	High	Moderate
Mngmt support	Moderate	High	*Low*	Moderate
Constit support	*Low*	High	Moderate	Moderate
Timeliness	*Low*	High	Moderate	Moderate
Relevance	*Low*	High	Moderate	Moderate
Info quality	*Low*	High	Satisfactory	Satisfactory

Last Updated by Pat Smith
On (date)   July 12, 2005

TOOL C2.4
Document assumptions and uncertainty

The goal of this step is to discuss how assumptions about future developments might drive success or failure of each option. To ensure that everyone is aware of the quality of the information used in the scorecard and the related conclusions, a list should be provided of the kinds of uncertainties and assumptions made in completing the scorecard along with the scorecard outputs. Some examples might include:

- Upcoming election may mean change of players.
- The organization's audited 2004 expenditures were not available at the time of this assessment. However, the final audited figures would not have significant impacts on the analysis conducted.
- A customer survey has not been completed since 2003, but initial data indicate that significant demographic changes may be occurring in certain service areas. The risk perceptions and the communication needs and preferences of the new residents were not known at the time of this assessment. This new population could significantly affect what risk communication strategies would be most appropriate.

Last Updated by___________________
On (date) ___________________

TOOL C3.2
Evaluation

The changes in key indicators over time can provide important means for demonstrating positive returns on risk communication program investments. In the step, participants should begin developing a list of key indicators, as well as evaluation measures for both the short and long-term. Some examples include:

Key indicators

1. Increased consumer support for infrastructure and improvements
2. Reduced rates of specific adverse health outcomes

Process measures (short-term)

1. Constituency support: number of people who attend public meetings relating to proposed water quality standards
2. Timeliness: how long it takes for information to be developed and distributed (e.g., CCR or public information brochure).

Impact measures (longer term)

1. Relevance to consumer needs: customer surveys or website comments
2. Quality of information: number and type of calls to customer service response centers

GLOSSARY

Contaminant A chemical, microorganism, or other substance that renders water unsafe or unusable.

Decision analysis A systematic, stepwise approach for identifying and assessing the relative value of different decision options. This approach uses formal tools; focuses the participants on defining the problem, and stating objectives and measures; and involves facilitated processes to produce informed judgments for addressing the problem.

Demographic Vital statistics of a population (e.g., gender, size, ethnicity)

Emerging Previously unidentified, not yet linked to adverse impacts, or not previously shown to occur in a given region.

Focus group A qualitative research method where 4-12 participants are guided through a specific topic by facilitators.

Goals Broad statements of what outcomes the group wants to achieve in the long-term by addressing the strategic issues.

Legitimacy A generalized perception or assumption that the actions of an entity are desirable, proper, or appropriate within some socially constructed system of norms, values, beliefs, and definitions. (See Suchman 1995.)

Power The extent to which a party has or can gain coercive, utilitarian or normative means to impose its will in the relationship. (See Etzioni 1964.)

Risk communication An interactive process of exchange of information and opinion among individuals, groups, and institutions; often involves multiple messages about the nature of risk or expressing concerns, opinions, or reactions to risk messages or to legal and institutional arrangements for risk management. (See NRC 1989.)

Scenario A hypothetical sequence of events constructed for the purpose of focusing attention on causal processes and decision-points. (See Kahn and Wiener 1967 cited in Cooke 1991).

Stakeholder Any group or individual who can affect or is affected by the achievement of the organization's objectives. (See Freeman 1984.)

Stakeholder dialogue A process whereby communicators define stakeholders around an issue, learn what the priorities are for those stakeholders, and developing communication strategies that address the interests and priorities of stakeholders.

Strategic issues Fundamental policy choices or critical challenges that must be addressed for a community [or collaboration] to achieve its vision. (See http://mapp.naccho.org.)

Strategies Systematic processes comprised of plans, methods, and a series of actions designed to balance issues and produce specific results.

Tabletop exercise A facilitated method for analyzing decision-making processes under various scenarios, conducted in an informal and stress-free environment. Participants use independently developed procedures and tools to conduct the exercise and evaluate both the experience and results.

Urgency The degree to which stakeholder claims call for immediate attention, under two conditions: (1) when the claim or relationship is time-sensitive, and (2) when that relationship or claim is important or critical to the stakeholder.

REFERENCES

ACPM (American College of Preventive Medicine). 2003. Waterborne Disease and Water Pollution: What Every Physician in Your Community Needs to Know [Online]. Available: <ttp://www.acpm.org/ehealth/waterborne.htm> [cited August 2003]

Anderson, E., and S.D. Jap. 2005. The Dark Side of Close Relationships. *MIT Sloan Management Review*, 46(3):75–83.

APHA (American Public Health Association). 2001. Public Health Code of Ethics [Online]. Available: <http://www.apha.org>. [cited March 28, 2002]

Argyris, C. 2001. Interpersonal Barriers to Decision Making. In: *Harvard Business Review on Decision Making*. Edited by P.F. Drucker, J. Hammond, R. Keeney, H. Raiffa, and A.M. Hayashi. Boston, Mass.: Harvard Business School Press.

Ashford, D.A., R.M. Kaiser, M.E. Bales, K. Shutt, A. Patrawall, A. McShan, J.W. Tappero, B.A. Perkins, and A.L. Dannenberg. 2003. Planning Against Biological Terrorism: Lessons From Outbreak Investigations. *Emerging Infectious Diseases*, 9(5):515–519.

AWWA (American Water Works Association). 2003. Fact Sheet: Fluoridation [Online]. Available: <http://www.awwa.org/Advocacy/pressroom/fluoride.cfm>. [cited August 2003]

Balbus, J.M., and M.A. Embrey. 2000. *Drinking Water and Disease: What Health Care Providers Should Know*. Washington, D.C.: Physicians for Social Responsibility.

Beierle, T.C., and J. Cayford. 2002. *Democracy in Practice: Public Participation in Environmental Decisions*. Washington, D.C.: Resources for the Future.

Blomqvist, K., P. Hurmelinna, and R. Seppaen. 2005. Playing the Collaboration Game Right—Balancing Trust and Contracting. *Technovation*, 25:497–504.

Brun, W. 1994. Risk Perception. In *Subjective Probability*. Edited by G. Wright and P. Ayton. New York: John Wiley and Sons.

CDC (Centers for Disease Control and Prevention). 1995. Assessing the Public Health Threat Associated with Waterborne Cryptosporidiosis: Report of a Workshop. *Morbidity & Mortality Weekly Report*, 44(RR-6):1–19.

CDC (Centers for Disease Control and Prevention). 1997. *Cryptosporidium and Water: A Public Health Handbook*. Atlanta, Ga.: Working Group on Waterborne Cryptosporidiosis. Atlanta, Ga.: CDC.

CDC (Centers for Disease Control and Prevention). 1999. Achievements in Public Health, 1900–1999: Fluoridation of Drinking Water to Prevent Dental Caries. *Morbidity & Mortality Weekly Report*, 48(41):933–940.

CDC (Centers for Disease Control and Prevention). 2001. Updated Guidelines for Evaluating Public Health Surveillance Systems: Recommendations from the Guidelines Working Group. *Morbidity & Mortality Weekly Report*, 50(RR-13):1–35.

CDC (Centers for Disease Control and Prevention). 2002. Surveillance for Waterborne-Disease Outbreaks–United States, 1999–2000. *Morbidity & Mortality Weekly Report*, 51(SS-08):1–28.

Chang, M., M.K. Glynn, and S.L. Groseclose. 2003. Endemic, Notifiable Bioterrorism-related Diseases, United States, 1992–1999. *Emerging Infectious Diseases,* 9(5):556–564.

Claiborne, N., and H.A. Lawson. 2005. An Intervention Framework for Collaboration. *Families in Society: The Journal of Contemporary Social Services,* 86(1):93–101.

Cooke, R.M. 1991. *Experts in Uncertainty: Opinion and Subjective Probability in Science*. New York: Oxford University Press.

CSA (Canadian Standards Association). 1997. *Q850 Risk Management: Guideline For Decision-Makers*. Toronto, Ont.: CSA.

Davidson, R.A., and R. Waddell. 2005. A Historical Overview of Interdisciplinary Family Health: A Community-based Interprofessional Health Professions Course. *Academic Medicine,* 80(4):334–338.

Driscoll, C., and Starik, M. 2004. The Primordial Stakeholder: Advancing the Conceptual Consideration of Stakeholder Status for the Natural Environment. *Journal of Business Ethics*, 49(1):55–73.

Drucker, P. 2001. The Effective Decision. In: *Harvard Business Review on Decision Making*. Edited by P.F. Drucker, J. Hammond, R. Keeney, H. Raiffa, and A.M. Hayashi. Boston, Mass.: Harvard Business School Press.

Etzioni, A. 1964. *Modern Organizations*. Englewood Cliffs, N.J.: Prentice-Hall.

Etzioni, A. 2001. Humble Decision Making. In: *Harvard Business Review on Decision Making*. Edited by P.F. Drucker, J. Hammond, R. Keeney, H. Raiffa, and A.M. Hayashi. Boston, Mass.: Harvard Business School Press.

Freeman, R.E. 1984. *Strategic Management: A Stakeholder Approach*. Boston, Mass.: Pitman.

Fuqua, J., D. Stokols, J. Gress, K. Phillips, and R. Harvey. 2004. Transdisciplinary Collaboration as a Basis for Enhancing the Science and Prevention of Substance Use and "Abuse." *Substance Use & Misuse*, 39(10-12):1457–1514.

Gorlin, R.A., ed. 1994. *Codes of Professional Responsibility.* 3rd ed. Washington, D.C.: Bureau of National Affairs.

Hammond, J.S., R.L. Keeney, and H. Raiffa. 2001. The Hidden Traps in Decision Making. In: *Harvard Business Review on Decision Making*. Edited by P.F. Drucker, J. Hammond, R. Keeney, H. Raiffa, and A.M. Hayashi. Boston, Mass.: Harvard Business School Press.

Hyde, J.S., W. DeJong, and S. Ratzan. 2000. Using Formative Research in the Risk Communication Process: Developing Consumer Confidence Reports – Results From Massachusetts. Presented at American Public Health Association Annual Meeting, Boston, Mass., November 12–16, 2000.

IOM (Institute of Medicine). 1988. *The Future of Public Health.* Washington, D.C.: National Academy Press.

Israel, B.A., A.J. Schulz, E.A. Parker, and A.B. Becker. 1998. Review of Community-Based Research: Assessing Partnership Approaches to Improve Public Health. *Annual Reviews of Public Health*, 19:173–202.

Jaeger, C.C., R. Ortwin, R.A. Rosa, and T. Webler. 2001. *Risk, Uncertainty, and Rational Action.* London: Earthscan Publications.

Jasuja, G.K., C.-P. Chou, K. Bernstein, E. Wang, M. McClure, and M.A. Pentz. 2005. Using Structural Characteristics of Community Coalitions to Predict Progress in Adopting Evidence-Based Prevention Programs. *Evaluation and Program Planning,* 28:173–184.

Johnson, B.B. 2000. Utility Customers' Views of the "Consumer Confidence Report" of Drinking Water Quality. *Risk: Health, Safety & Environment*, 11:309–328.

Karoly, L.A., M.R. Kilburn, J.H. Bigelow, J.P. Caulkins, and J.S. Cannon. 2001. *Assessing Costs and Benefits of Early Childhood Intervention Programs: Overview and Application to the Starting Early Starting Smart Program.* Santa Monica, Calif.: RAND Corporation.

Keeney, R., and H. Raiffa. 1993. *Decisions with Multiple Objectives*. New York: Cambridge University Press.

Krimsky, S., and D. Golding, eds. 1992. *Social Theories of Risk*. Westport, Conn.: Praeger.

Lacher, D., E. Nelson, W. Bylsma, and R. Spena. 2000. Computer Use and the Needs of Internists: A Survey of Members of the American College of Physicians and the American Society of Internal Medicine. In *Proceedings of the American Medical Informatics Association Symposium*. Philadelphia, Pa.: Hanley and Belfus.

LaPorte, R.E., D.J. McCary, T.J. Songer, G. Bruno, and N. Tajima. 1993. Disease Monitoring. *The Lancet*, 341:1416.

Lazo, J.K., J.L. Pratt, C.N. Herrick, R.E. Hurd, and J. Rambo. 2004. *Understanding and Enhancing the Impact of Consumer Confidence Reports*. Denver, Colo.: AwwaRF.

Lim, G., H. Ahn, and H. Lee. 2005. Formulating Strategies for Stakeholder Management: A Case-Based Reasoning Approach. *Expert Systems With Applications*, 28(4):831–840.

Lindgren, M., and H. Bandhold. 2003. *Scenario Planning: The Link Between Future and Strategy*. London: Palgrave Macmillan.

Magee, M. 2003. Qualities of Enduring Cross-Sector Partnerships in Public Health. *American Journal of Surgery*, 185(1):26–29.

McCarty, K., J. Swallow, R. Vanderslice, and W.S. Combs. 2000. Water Systems to Report Drinking Water Quality to All Customers: How Can Health Professionals Prepare for the Questions That These Reports Will Generate? *Medicine and Health Rhode Island*, 83:140–143.

Mitchell, R.K., B.R. Agle, and D.J. Wood. 1997. Toward a Theory of Stakeholder Identification and Salience: Defining the Principle of Who and What Really Counts. *Academy of Management Review*, 22(4):853–866.

Mobley, J., E. Tatham, K. Reinhardt, and C. Tatham. 2005. *Message Management: Effective Communications*. Denver, Colo.: AwwaRF.

Morgan, M.G., B. Fischhoff, A. Bostrom, and C.J. Atman. 2002. *Risk Communication: A Mental Models Approach*. New York: Cambridge University Press.

NAS (National Academy of Sciences). 2004. Benefit-Cost Analysis in Water Project Planning. In *Analytical Methods and Approaches for Water Resources Project Planning*. Washington, D.C.: National Academy Press.

NDWAC (National Drinking Water Advisory Council). 2005. *Water Security Working Group Findings*. Washington, D.C.: USEPA.

Nebraska Cooperative Extension. 2002. Drinking Water: Chloramines Water Disinfection in Omaha Metropolitan Utilities District [Online]. Available: <http://www.ianr.unl.edu/pubs/water/nf505.htm>. [cited August 2003]

NEETF (National Environmental Education and Training Foundation). 1998. *The National Report Card on Safe Drinking Water Knowledge, Attitudes and Behaviors*. Washington, D.C.: NEETF.

NEETF (National Environmental Education and Training Foundation). 1999. *The National Report Card on Safe Drinking Water Knowledge, Attitudes and Behaviors*. Washington, D.C.: NEETF.

NRC (National Research Council). 1983. *Risk Assessment in the Federal Government: Managing the Process*. Washington, D.C.: National Academy Press.

NRC (National Research Council). 1989. *Improving Risk Communication*. Washington, D.C.: National Academy Press.

NRC (National Research Council). 1996a. *Health Communities: New Partnerships for the Future of Public Health*. Washington, D.C.: National Academy Press.

NRC (National Research Council). 1996b. *Understanding Risk: Informing Decisions in a Democratic Society*. Washington, D.C.: National Academy Press.

NSWHD (New South Wales Health Department). 2001. Memorandum of Understanding Between NSW Health and Sydney Water Corporation [Online]. Available: <http://www.health.nsw.gov.au>. [cited June 9, 2005]

O'Connell, L.L., C.U. Stephens, M. Betz, J.M. Shepard, and J.R. Hendry. 2005. An Organizational Field Approach to Corporate Rationality: The Role of Stakeholder Activism. *Business Ethics Quarterly*, 15(1):93–111.

O'Connor, D., and C.T. Gates. 2000. Toward a Healthy Democracy. *Public Health Reports*, 143:157–160.

Ogilvy, J.A. 2002. *Creating Better Futures: Scenario Planning as a Tool for a Better Tomorrow*. New York: Oxford University Press.

OMB (United States Office of Management and Budget), Office of Information and Regulatory Affairs. 2003. *Informing Regulatory Decisions: 2003 Report to Congress on the Costs and Benefits of Federal Regulations and Unfunded Mandates on State, Local, and Tribal Entities*. Washington, D.C.: U.S. OMB.

Parkin, R. 2005. Strategic Risk Communication: The Challenge of Emerging Drinking Water Contaminants. *Risk Policy Report*, 12(7):14–16.

Pettiti, D. 2000. *Meta-Analysis, Decision Analysis and Cost-Benefit Analysis*. New York: Oxford University Press.

Pike, W., B. Yarnal, A.M. MacEachren, M. Gahegan, and C. Yu. 2005. Retooling Collaboration: A Vision for Environmental Change Research. *Environment*, 47(2):9–21.

Pluye, P., L. Potvin, J.-L. Denis, J. Pelletier, and C. Mannoni. 2005. Program Sustainability Begins With First Events. *Evaluation and Program Planning*, 28:123–137.

Porter, J., and E.L. Baker. 2005. The Management Moment: Partnership Essential. *Journal of Public Health Management Practice*, 11(2):174–177.

Rhoten, D. 2004. Navigating Interdisciplinary Collaboration. Presented at the Annual Meeting of the Society of Risk Analysis, Palm Springs, Calif., December 5–8, 2004.

Rice, P.L., and D. Ezzy. 1999. *Qualitative Research Methods: A Health Focus*. South Melbourne, Australia: Oxford University Press.

Ritchie, J., and J. Lewis, eds. 2003. *Qualitative Research Practice: A Guide for Social Science Students and Researchers*. London: Sage Publications.

Rosen, G. 1993. *The History of Public Health*. Baltimore, Md.: Johns Hopkins University Press.

SDWA (Safe Drinking Water Act) Amendments of 1996. 42 *U.S. Code*. Sec. 300f–300j-26.

Slovic, P., ed. 2000. *The Perceptions of Risk*. London: Earthscan Publications.

Squires, S.G., K.J. Aronson, R.S. Remis, and J.R. Hoey. 1998. Improved Disease Reporting: A Randomized Trial of Physicians. *Canadian Journal of Public Health*, 89(1):66–69.

Stevens, R. 2001. Public Roles for the Medical Profession in the United States: Beyond Theories of Decline and Fall. *The Milbank Quarterly*, 79(3):327–353.

Stokols, D. 1995. Translating Social Ecological Theory Into Guidelines for Community Health Promotion. *American Journal of Health Promotion*, 10(4):282–298.

Stokols, D., J. Fuqua, J. Gress, R. Harvey, K. Phillips, L. Baezconde-Garbanati, J. Unger, P. Palmer, M.A. Clark, S.M. Colby, G. Morgan, and W. Trochim. 2003. Evaluating Transdisciplinary Science. *Nicotine & Tobacco Research*, 5(Supplement 1):S21–S39.

Suchman, M.C. 1995. Managing Legitimacy: Strategic and Institutional Approaches. *Academy of Management Review*, 20:571–610.

USEPA (United States Environmental Protection Agency). 1998. National Primary Drinking Water Regulations: Consumer Confidence Reports. *Federal Register*, 63(160):44511–44536.

USEPA (United States Environmental Protection Agency). 2003. Analysis and Findings of the Gallup Organization's Drinking Water Customer Satisfaction Survey. Washington, D.C.: USEPA, Office of Ground Water and Drinking Water.

von Winterfeldt, D. 1992. Expert Knowledge and Public Values in Risk Management: The Role of Decision Analysis. In: *Social Theories of Risk*. Edited by S. Krimsky and D. Golding. Westport, Conn.: Praeger.

ABBREVIATIONS

ACPM	American College of Preventive Medicine
AOEC	Association of Occupational and Environmental Clinics
AWWA	American Water Works Association
AwwaRF	Awwa Research Foundation
BWA	boil water alert
CCR	Consumer Confidence Report
CDC	United States Centers for Disease Control and Prevention
DEQ	Department of Environmental Quality
DNR	Department of Natural Resources
ERP	emergency response plan
GWU	The George Washington University
HCP	health care provider
IRB	Institutional Review Board
LPHA	local public health agency
MAPP	Mobilizing for Action through Planning and Partnerships
MOU	memorandum of understanding
MTBE	methyl tertiary butyl ether
NACCHO	National Association of City and County Health Officials
NDWAC	National Drinking Water Advisory Council
PAC	project advisory committee
PDPH	Philadelphia Department of Public Health
PHA	public health agency
PI	principal investigator
PIO	public information officer
S/LPHA	state and local public health agencies
SPHA	state public health agency
USEPA	United States Environmental Protection Agency

IWA Publishing's authorised EU representative for General Product Safety Regulations is Diane D'Arras, 15 rue Duret, 75116 Paris, France, e-mail: safety@iwap.co.uk.

Printed and bound by CPI Group (UK) Ltd, Croydon, CR0 4YY
18/09/2026
02218938-0003